Global Intelligence and Human Development

Global Intelligence and Human Development

Toward an Ecology of Global Learning

Mihai I. Spariosu

The MIT Press
Cambridge, Massachusetts
London, England

MIT Press books may be purchased at special quantity discounts for business or sales promotional use. For information, please email special_sales@mitpress.mit.edu or write to Special Sales Department, The MIT Press, 5 Cambridge Center, Cambridge, MA 02142.

This book was set in Sabon on 3B2 by Asco Typesetters, Hong Kong, and was printed and bound in the United States of America.

Library of Congress Cataloging-in-Publication Data

Spariosu, Mihai.
Global intelligence and human development: toward an ecology of global learning / Mihai I. Spariosu.
p. cm.
Includes bibliographical references and index.
ISBN 0-262-19511-9 (hc. : alk. paper) — ISBN 0-262-69316-X (pbk. : alk. paper)
1. International education. 2. Globalization. 3. Multicultural education.
I. Title.
LC1090.S65 2005
370.116—dc22 2004042613

10 9 8 7 6 5 4 3 2 1

For Diana, Ana Maria, and Miguel Antonio,
with all my love

What shall I say, O' Muslims, I know not myself
I am neither a Christian, nor a Jew, nor a Zoroastrian, nor a Muslim
Neither of the East, nor of the West, nor of the desert, nor of the sea
Neither from the land, nor of the sky
Neither of the earth, nor of water, nor of wind, nor of fire
Neither of the high, nor of low, nor of space, nor of time
Neither an Indian, nor Chinese, nor Bulghar, nor Saksin
Neither of Iraq, nor of Khorasan
Neither of this world, nor of the next, nor of paradise, nor of hell
Neither of Adam, nor of Eve.
My place is the placeless, my sign is the signless
There is neither a body nor a soul
For I am of the Beloved.

—Rumi, thirteenth-century Sufi poet (translated from the Persian by Majid Tehranian)

Contents

Acknowledgments

I would like to express my gratitude to friends, colleagues, and students, from many parts of the world, that have contributed directly or indirectly to this book and are too numerous to mention here by name. I am particularly grateful to Sorin Antohi, Ronald Bogue, Mikhail Epstein, Jacques de Pablo Lacoste, Jeannie Lum, Mike Featherstone, Rose-Lynn Fisher, Wlad Godzich, Wolfgang Iser, Giuseppe Mazzotta, Riel Miller, Virgil Nemoianu, Brian Rosborough, Hardy F. Schloer, and Zhang Xinhua who read all or parts of the manuscript and gave me their valuable comments and suggestions.

Additionally, Rose-Lynn Fisher's serendipitous visit to the Book Expo of America in Los Angeles, one day in the spring of 2003, has triggered the happy, intercontinental butterfly effect that led to the publication of this book with The MIT Press. I am equally grateful to the anonymous readers, as well as to Clay Morgan, Senior Editor at this press, for their excellent comments and suggestions that led to extensive revisions of an earlier draft of the manuscript. I would also like to thank Majid Tehranian for drawing my attention to and providing the translation of the Rumi poem that I have used in the epigraph.

Parts of the introduction to this book are based on an earlier essay, "Global Intelligence and Intercultural Studies," published by *Turnrow* magazine in 2003. The discussion of E. O. Wilson's *Consilience* in chapter 3 is also based on an earlier essay, published in a collective volume, *The Finer Grain*, by Indiana University Press in 2003.

Last, but certainly not least, I would like to thank my wife Diana Santiago and my children Ana Maria and Michael Anthony, who have put up with my long working hours, frequent traveling, and absentee fatherhood. They have taught me, above all, that intercultural communication and understanding, as well as unconditional love, begin at home.

Introduction: What Is Global Intelligence?

The vertiginous pace of global change in recent decades has led many scholars and practitioners to conclude that humanity stands at the threshold of a new era. They partly attribute this development to the veritable explosion of new technologies and technosciences. Information and communication technology (ICT) enables a significant number of individuals throughout the world to communicate with each other almost instantaneously. The personal computer and the World Wide Web have created a global communication medium through which each participant becomes a node in an easily accessible and readily affordable informational network. Under the impact of ICT, our ways of acquiring, transmitting, and using knowledge are said to undergo changes as well: cognitive paradigms appear to be shifting from linear, binary forms, which favor disciplinary and compartmentalized modes of knowledge, to nonlinear, holistic forms, which favor transdisciplinary, integrative modes. World economies and governance are consequently seen as driven by information or knowledge that is intended, at least in principle, to be accessible to and shared by all.

According to our futurologists, moreover, advanced energy and materials technology will soon allow humanity to embark on global projects that have so far been the domain of science fiction, such as iceberg towing for arid zone irrigation, ocean mining, waste disposal into the earth's mantle, earthquake prevention, weather modification, fail-safe nuclear power plants, automated farming and animal husbandry, integrated water supply systems on a continental scale, robotic assists for people, nanoscale products and systems, planning for terraforming, and so on. Finally, genetics, medical science, and brain technology, experts tell us, seem on the brink of decoding and duplicating the processes of life itself, thereby helping humans live a longer, healthier, and physically more satisfying existence, if not changing the very definition of life on earth.

But global changes are also perceived to be of a sociopolitical and cultural nature. The recent collapse of a large number of right-wing and left-wing dictatorships around the planet appears to show a renewed desire on the part of world communities to move toward various forms of open society. The laissez-faire ideology of the marketplace seems to have replaced authoritarian political ideologies in different parts of the world, and liberal democracy seems to have become the preferred mode of government—if not in fact, at least as a stated political ideal. Global free markets and finance seem, in turn, to render the old national borders more flexible and permeable, bringing unexpected economic development and an abundance of services and material goods to places that modernity has hitherto bypassed. Above all, the spirit of the age appears to be moving toward an awareness of our globe as a single place, as a great plurality of culturally diverse, yet interdependent localities. All of these worldwide developments have been variously hailed as the New Age, New World Order, End of History, modernity at large, postmodernity and, most recently, globalization.

Irrespective of the economic, political, cultural, or spiritual labels for our age, most thinkers agree that human problems have now become global, that is, highly complex, nonlinear, transdisciplinary, and transnational in nature. These problems can therefore be addressed only through a concerted effort on the part of our world communities. But how can one achieve such global unity of purpose in a world that continues to be divided by social and economic inequity, ideological, political, and ethnic strife, and intercultural misunderstanding and distrust? Can, for example, a global free market, supported by advanced technosciences accessible to all, take care of these problems in and by itself, as many high priests of neoliberal commerce claim? Even such successful global financiers as George Soros conclusively argue that this cannot be the case, because so-called free markets are often driven by greed and material profit, rather than by high-minded values, and therefore need various correctives (Soros 1998).

But who is to devise and apply such correctives? The new global paradigm seems equally to imply that no single political, economic, or social force is in control and, therefore, that there is no longer any possibility of effective individual or collective action. The conventional international mechanisms and organizations that are currently in place, such as the United Nations, the World Trade Organization, the International Monetary Fund, the World Bank, the World Court and other international courts of law, seem inadequate in dealing with these new global developments and are in dire need of extensive reforms, although there is little consensus about the direction in which these reforms should go.

There are also those thinkers and practitioners who see the present globalizing trends as precipitating humankind into Armageddon. According to these mostly Marxist and left-wing scholars, but also extreme, religious and other millenarian, right-wing figures, the forces of evil assume the artful guises of neoliberal capitalism and/or of an increasingly cynical and unbridled will to power, well on its way to creating a new kind of global hegemony or empire. These prophets of doom certainly do not lack evidence for their bleak views, citing the increasing poverty of world populations, which corresponds to an ever greater concentration of the world's wealth in fewer and fewer hands; the worrisome spread of ethnic and religious violent conflicts and political terrorism; and a rapid depletion of our planet's biodiversity and natural resources.

None of these prophets, however, can offer any innovative solutions to what they perceive as an ever worsening global plight. They invoke either the same old threats of fire and brimstone, or the same old activist remedies such as resistance to capitalist exploitation, concerted trade unionist action, state and suprastate regulations to curb the greedy speculations of global financial markets, and so on. More fancifully, some of them posit the advent of latter-day Nietzschean barbarians who will topple the new empire and, ironically, usher in yet another Christian Kingdom of Love.[1] Other thinkers, who realize that these antiquated or queer remedies are part of the problem, rather than the solution, simply wring their hands in despair or contend themselves with searing critiques of the new global (dis-)order, thereby contributing to the general despondency and disarray of a large section of the global intellectual community.

All of these developments seem to indicate that our planet might indeed be on the brink of a radical paradigmatic shift. This shift may or may not occur, whether in this generation or the next, independent of the multiplying prophesies of impending doom or commercial paradise. But if it does, it will be brought about neither by our technosciences, nor by neoliberal market forces, nor by "freedom," "resistance," and "fundamentalist" movements, nor by a New World Order, nor by new barbarians. Rather, it will be brought about by each of us. Far from being helpless spectators, we human beings are directly implicated in what the immediate future holds. Each of us is responsible for what will happen next, whether we know it or not, and whether we like it or not. Hence, global awareness means understanding not only the interdependence of all localities within the global framework, but also the enhanced individual responsibilities that result from it.

There are neither ready-made solutions nor quick fixes to world problems, but what we need to work on collectively, throughout the planet, is a change in our modes of relating to each other and to our natural habitat, which are proving to be less and less sustainable in the current global circumstance. Yet, how can we effect changes in our behavior patterns through means other than technological, economic, political, or religious? Would social and human engineering not be the obvious way, after all, given the advanced state of our technologies? The catastrophic outcome of the extensive social and human experiments conducted on large populations throughout the world during the past century should, however, give us pause before we embark on such ill-fated adventures in the future.

One way—perhaps the only lasting way—in which human nature can change is through (self-)education, as Plato was fully aware when he founded his Academy and wrote the *Republic* and the *Laws*. If the twenty-first century is to be the century of education, as the proponents of the "knowledge economy" tell us, then we should take them at their word: as we continue seeking sustainable solutions to world problems, we should concentrate our efforts on educating ourselves, our children, and our grandchildren. If we wish to develop a different kind of global mindset, we need to create learning environments that will help us rise to this seemingly impossible challenge, without repeating the errors of the past, including those of the Platonic Academy.

Our global pundits, whether on the right or the left, seem to connect human progress primarily with material development. Most worldwide statistics and indicators are economic in nature, measuring human happiness by what an individual or a social group has, rather than by what they are. Thus, we have presently divided the world into "developed," "underdeveloped," and "developing" societies. But if we truly wish to change our global paradigms, then we need to change the focus of our worldwide efforts from social and economic development to human self-development. From the standpoint of the latter, there are no developed or underdeveloped societies, but only developing ones. It is this kind of development that in the end will help us solve our practical problems, including world hunger, poverty, and violence, and will turn the earth into a welcoming and nurturing home for all of its inhabitants, human and nonhuman.

I am not sure that we have quite as yet reached a juncture, in the collective life of humanity, at which we are ready to undertake the radical worldwide reforms needed to unleash our true human potentialities, even though many of these reforms and their promise are encoded in the collective wisdom of our world civilizations. I do believe, however, that the next two or three generations can at least prepare the way for such reforms, that is, prepare the conditions that will allow their worldwide emergence and implementation. It is for this reason that humanity's common project for the next few decades should be an educational one. And its main objective should be the creation of local–global learning environments throughout the world, which will open us toward our true human possibilities.

The importance of global education has increasingly come to public attention in the wake of recent world events. But our educators and other practitioners in the field of learning and research have, at least so far, stopped short of adopting a genuinely global approach to world education. For example, a white paper on "Beyond September 11: A Comprehensive National Policy on International Education," generated by the American Council on Education (ACE) and signed by thirty-three other U.S. higher education organizations, calls for extensive reforms in the world of North American higher education, especially in terms of what it calls *global competence*. The paper defines global competence as "in-depth knowledge required for interpreting information affecting national security, the skills and understanding that foster improved relations with all regions of the world; ... foreign language proficiency and an ability to function effectively in other cultural environments and value systems, whether conducting business, implementing international development projects, or carrying out diplomatic missions" (ACE 2002, p. 1). The paper also calls for the creation of "global experts in foreign languages, cultures and political, economic and social systems throughout the world" (ibid., p. 2).

Global competence and expertise are certainly very important talents and skills to be developed in our national citizenry and workforce. But, for the ACE paper, the operative word remains "national." Although it deals with global issues, this paper adopts a national or an international, rather than a global perspective on such issues. A global approach would take into consideration not only the perceived national or "local" interests of the United States or any other country or region. Of course, those local interests are extremely important, and genuine global practitioners will neglect them only at their peril. But such global practitioners would also look beyond what might turn out to be short-term and limited national interest to long-range interests serving the entire global community. From this global perspective, the concept of national interest itself may gain a new dimension and be redefined, in a larger reference frame, as that which ultimately is in the best interest of and benefits all nations and cultures.

Therefore, the global learning and research environments for which I plead in the present book are primarily meant to advance *global intelligence*, of which global competence and global expertise might be side benefits. I define global intelligence as the ability to understand, respond to, and work toward what is in the best interest of and will benefit all human beings and all other life on our planet. This kind of responsive understanding and action can only emerge from continuing intercultural research, dialogue, negotiation, and mutual cooperation; in other words, it is interactive, and no single national or supranational instance or authority can predetermine its outcome. Thus, global intelligence, or intercultural responsive understanding and action, is what contemporary nonlinear science calls an emergent phenomenon, involving lifelong learning processes.

Global intelligence also implies what Theodore Roszak, in his fine book *Unfinished Animal: The Aquarian Frontier and the Evolution of Consciousness* (1975), terms "spiritual intelligence." Roszak carries out a devastating critique of the "consciousness circuit" of the New Age, with its "lethal swamp of paranormal entertainments, facile therapeutic tricks, authoritarian guru tricks, demonic subversions" (Roszak 1975, p. 13). He nevertheless understands the great importance of the consciousness circuit as a source of human development and renewal, if it is informed by spiritual intelligence, which he defines as the "power to tell the greater from the lesser reality, the sacred paradigm from its copies and secular counterfeits" (ibid.). According to Roszak, spiritual intelligence can be found in the sacred traditions of the world, "in those ancient springs of visionary knowledge which are the source of the mystic and occult schools, and from which we draw our entire repertory of transcendent symbolism and transcendental insight" (ibid.).

Variously called the "perennial wisdom," the "secret doctrine," or the "old gnosis," one can find traces of this ancient visionary knowledge practically all over the planet. It is present, for example, in the Rumi poem that I have used as an epigraph to the present book and that I shall discuss in detail in chapter 4 below. It could, indeed, become an important source of global learning, but not before its principles and practices are also submitted to an extensive intercultural research and dialogue. It is imperative, moreover, to evaluate this source of wisdom not only in relation to its goals or ends, but also in relation to its means, that is, its past and present institutional frameworks and practices. If one examines the various mystical schools and secret societies, including Plato's Academy, a wide gap between their generous humanitarian ends and inapposite means will soon come into view. Most of them encrypt ancient wisdom in elitist secret codes and hierarchical institutional frames, thereby obscuring or even perverting its global, ecumenical intent. But, as Lao Tzu, Gautama Buddha, Pythagoras, Jesus of Nazareth, Rumi, and other teachers of humanity have pointed out, there should be nothing esoteric or mystical (i.e., hidden) in this simple and transparent wisdom that should be readily available to, and practicable by, every human being.

Another reason why most of these occult or esoteric schools have so far failed to change the course of human development is because they have largely misunderstood spiritual practice as a discipline, in all the senses of that term, directed mostly against the material world. By stressing mind or spirit over body or matter, often "mortifying" or repressing the latter, they have neglected the underlying unity of the human and other worlds, in which spirit and matter are complementary and interdependent, engaged in a relationship of causal reciprocity or mutual causality. Esoteric schools have more often than not essentialized these complementary, interior and exterior, dimensions of reality, turning them into conflictive, asymmetrical binary oppositions. Consequently, one dimension or the other has gained the upper hand in human history, with self-destructive excesses on both sides. Although Roszak is quite right in deploring the excessive materialism that prevails in North American and, for that matter, in many other contemporary Western and non-Western cultures, freestanding spirituality is hardly an answer to it, either. Roszak himself instructively and wittily exposes the spiritualist excesses of contemporary counterculture, which often ends up enmeshed, just like its older avatars, in the very materialism it wishes to eradicate.

Global intelligence, then, would imply both spiritual and material intelligence. It would regard human and other forms of consciousness as interior dimensions of the physical world, not subordinated to, but coextensive with it. It would see spirituality and materiality in terms of dependent co-arising (to use the Buddhist term for mutual or reciprocal causality that I shall discuss at length in chapter 4 below), and would distinguish them from spiritualism and materialism, with their linear notion of causality that has led each of them to regard itself as the "first cause" of the other.

Global intelligence would also imply what Peter Singer, in the wake of Victorian philosopher Henry Sidgwick (in the *Methods of Ethics*, 1874), calls "taking the point of view of the universe." From this viewpoint, "we can see that our own sufferings and pleasures are very like the sufferings and pleasures of others; and that there is no reason to give less consideration to the sufferings of others, just because they are 'other.' This remains true in whatever way 'otherness' is defined, as long as the capacity for suffering or pleasure remains" (Singer 1995, p. 222). Starting from the utilitarian premise of the primacy of pleasure and pain—which is only one side of Sidgwick's philosophy, however—Singer develops the principles of the "good life" on a global scale, in terms of what one might call an "enlightened utilitarianism," in which the notion of self-interest is expanded to embrace the whole sentient universe.

The practical global ethics that Singer proposes is equally present in the ancient traditions of wisdom, as he himself mentions, in the form of the Golden Rule. For example, Jesus says, "Love your neighbor as yourself"; Rabbi Hillel says, "What is hateful to you do not do to your neighbor"; the Indian epic *Mahabharata* says, "Let no man do to another that which would be repugnant to himself"; and Confucius similarly says, "What you do not want done to yourself, do not do to others" (ibid., p. 230). Singer, however, interprets the Golden Rule in a narrow, utilitarian way, as encouraging "equal consideration of interests" (ibid.). From the perspective of global intelligence, the Golden Rule goes well beyond such considerations, implying a completely different human mentality, based on peaceful cooperation, responsive understanding, and love of the other, instead of contest, conflict, and material self-interest. As such, the Golden Rule remains a basic principle of human interaction, within and outside a global reference frame.

In turn, Singer's enlightened utilitarianism of Victorian extraction is certainly an improvement over the unrelenting, utilitarian materialism of our "age of self-interest," as he calls it. Yet, he sets the bar too low in attempting to transvaluate our mainstream, utilitarian values: when all is said and done, he offers nothing but more utilitarianism. Unlike Nietzsche, for example, who traces the feelings of pleasure and pain directly to the will to power, Singer never explores the cultural dimensions of these feelings, nor does he consider redefining them in terms other than utilitarian ones.[2] He simply accepts the premise that these feelings are universal, physiological phenomena, as "natural" as eating, drinking, and sexual intercourse, to which they are routinely linked. (As if such universals would not vary a great deal from culture to culture and would not in turn receive their natural content from various cultural systems of values and beliefs.)

Singer's notions of the universe and of a universal perspective, moreover, remain well within the Cartesian, dualistic paradigm in which sentient and physical objects are independent of each other and should be treated differently, according to their higher or lower degrees of being. For example, he notes: "I shall use Sidgwick's phrase to refer to a point of view that is maximally all-embracing, while not attributing any kind of consciousness or other attitudes to the universe, or any part of it that is not a sentient being" (Singer 1995, p. 222). Here, again, he reduces Sidgwick's phrase to a narrow, strictly utilitarian interpretation, thus ignoring that Victorian thinker's intuitive side that goes beyond material considerations to religious sentiment and human desire for transcendence. Also, unlike the deep ecologists who embrace, say, the Gaia hypothesis, suggesting that our planet, far from consisting of "dead matter" on which life develops, is itself a very complex self-organizing system (see chapter 3 below), Singer ignores the complex feedback loops between the animate and the inanimate worlds, considering them only in the linear terms of how they might affect human material interests.

Global intelligence would, then, transcend the utilitarian and "rights" theories of contemporary analytic philosophers that can hardly provide sustainable ethical grounds for human interaction within a global reference frame, confined as they are to the Western rationalist tradition. Instead, it would imply a holistic ecological understanding of our universe, advocated, for example, by "deep" ecologists (Naess 1973), who distance themselves from "shallow" ecologists such as Singer by extending the concept of self-interest to include the entire planet, not just sentient beings. Other deep ecologists, such as David Orr (1993) and, in his wake, Fritjof Capra (1997), call this holistic understanding "ecological literacy." Orr stresses the importance of building ecological literacy through a system of education that integrates curriculum and teaching with one's situation in the physical and cultural world. At a minimum, an "ecologically literate" person should learn about the earth as a physical system, the relation between ecology and thermodynamics, the earth's "vital signs" that indicate its well-being or illness, the essentials of human ecology, the natural history of one's own region, and how to restore natural systems and build sustainable communities and economies.

In turn, for Capra, being ecologically literate, or "ecoliterate," means "understanding the principles of organization of ecological communities (i.e., ecosystems) and using those principles for creating sustainable human communities—including our educational communities, business communities, and political communities—so that the principles of ecology become manifest in them as principles of education, management, and politics" (Capra 1997, p. 289). The organizing principles of natural ecosystems that Capra lists include: interdependence, recycling of waste, use of solar energy, symbiotic partnerships, flexibility, and diversity.

Orr's and Capra's are certainly sound principles that should be conducive to global intelligence. We need, however, to reflect on the Western ideological baggage that the term "ecological literacy" carries with it and, consequently, on its larger, intercultural implications. This term invokes the asymmetrical binary opposition of literate and "illiterate," that is, oral, societies—a conflictive polarity that has its own long and troubled intercultural history. Ecology would undoubtedly want to turn away from this polarity and generate productive, rather than destructive, feedback loops between the two kinds of societies. Incidentally, traditional, small-scale, oral communities have often exhibited a much more responsible "ecoliterate" behavior toward their environments than many members of our "literate" Western communities have toward ours. So, we can learn from them in this respect as much as we can learn from "nature." On the other hand, we should not idealize these societies, as some anthropological and ecological schools do, but attempt to understand their systems of values and beliefs in their own terms and then engage them in a genuine, productive dialogue.

The term *ecoliteracy*, then, whereas it may serve as a useful political slogan within the Western world, may resonate negatively with other societies and consequently become counterproductive within a global reference frame. By contrast, *ecological awareness* or *attentiveness* would be more appropriate phrases, from the standpoint of global intelligence, for conveying our ecological message to other cultures, as well as to our own.

Another problem arises when Orr, Capra, and other Western-style ecologists encourage us to mimic or imitate natural processes. Thereby, no less than Singer, they preserve the linear distinction between nature and culture and overlook the nonlinear principle of mutual causality that ought to describe their reciprocal interrelation, just as it describes that of spirit and matter, or that of animate and inanimate objects. In this sense, there is no pristine nature that we ought to imitate, any more than there is a fallen culture that perverts it. When ecologists ask us to mimic nature, they simply ask us to follow the "ecosystems" idea of nature, that is, a certain Western scientific assumption about it, just as sociobiology, for instance, invokes another Western scientific notion of nature, the "war of all against all," to justify its particular approach.

We thus tend to see in our physical environment the same destructive or productive feedback loops that we produce in our scientific and cultural systems in general. In mimicking nature, we risk mimicking ourselves, because nature will ceaselessly reflect back to us our own approaches to it. It follows that, if we truly wish to change ourselves and nature, we need to start from our mentality, which will then elicit the desired, productive, feedback responses from nature as well. It is for these reasons that I shall, in parts I and II of the present study, propose the notion of resonance, rather than imitation, as being more appropriate and rewarding to develop in a global ecological environment.

One fascinating aspect of the relationship of nature and culture is the interaction of humans and machines. This is a very old issue in the Western tradition, going all the way back to Hellenic mythology, for example, to the myth of Daedalus the Artificer and his hubristic son, Icarus. But, in our age of digital computers and unprecedented explosion of information technology, it has become of utmost urgency. Indeed, some thinkers relate it directly to the development of global intelligence. For example, George B. Dyson, in his informative study, *Darwin among the Machines: The Evolution of Global Intelligence* (1997) thoroughly explores the thesis that human-built machines are potentially the next, more advanced, intelligent form of life on the evolutionary scale. Among other things, he traces the rich history of this idea from the industrial age to the digital age, from such visionaries as Samuel Butler and H. G. Wells to the present-day developers of artificial intelligence (AI), such as Marvin Minsky. Dyson persuasively argues that machines, no less than human cultures in general, are equally part of nature and, as such, subject to natural evolution.

For Dyson, global intelligence surpasses any specific individual, emerging from the interaction of all individuals. To support this view, Dyson invokes H. G. Wells's book, *The World Brain* (1938), in which Wells writes: "This new all-human cerebrum ... need not be concentrated in any one single place. It need not be vulnerable as a human head or a human heart is vulnerable. It can be reproduced exactly and fully, in Peru, China, Iceland, Central Africa and wherever else seems to afford an insurance against danger and interruption. It can have at once, the concentration of a craniate animal and the diffused vitality of an amoeba" (Dyson 1997, p. 11).

In turn, Dyson speculates that the self-organizing, rhizomic structure of the Internet might be the sign of a new global intelligence that is spontaneously emerging from the interaction of humans and computer networks. For him, the development of the Internet shows that humans will eventually be unable to control their "artificial" creations, the machines. In this respect, they are just like parents who can no longer control or guide their offspring once they come of age. Computers and their worldwide networks are slowly becoming part of natural evolution, in accordance with cooperative, symbiotic processes that Dyson, in the wake of some deep ecologists, believes have always driven such evolution.

Furthermore, some scholars advance the related claim that the new technologies will create a new type of sentient being, or the "metaman" (Stock 1993); a combination of human, computer, and other high-tech ware that will vastly exceed the physical and mental capabilities of *Homo sapiens*. To these scholars, the cyborgs and the bionic men and women of science fiction appear as an all-too-immediate reality. The question remains open, however, if such "metamen," no less than Dyson's self-organizing Internet, will also exceed the *ethical* capabilities of *Homo sapiens*. Indeed, will they ever be more than high-tech versions of Nietzsche's supermen, as they are largely portrayed in contemporary popular culture, especially in Hollywood "blockbusters"?[3]

I explore these issues at length in *Remapping Knowledge: Intercultural Studies for a Global Age*, a companion volume to the present one. Here I would simply like to point out, again, that global intelligence would involve not only a collective, but also an individual form of awareness and, therefore, responsibility, as H. G. Wells equally implies. Cooperation and symbiosis are not absolute values, moreover, but depend on the overall nature of the value system within which they are inscribed. Although I agree with Dyson that the interactions between humans and machines are part of natural evolution, I cannot emphasize enough that nature and culture cannot be separated, being inextricably woven within the same "web of life," to use Capra's felicitous phrase. Consequently, the evolutionary future of our machines depends on us, as much as our future depends on them.

I believe that we can, as human beings, have a decisive influence on the course that natural evolution will take. A new kind of global intelligence might well emerge from the symbiosis of humans and machines, but the creative or destructive nature of this intelligence will largely depend on us. As in the case of nature, machines will respond to or resonate with us in the same way that we approach them. In other words, we engage them in creative or destructive feedback loops, just as we do anything else within our universe.

Despite euphoric claims to the contrary, our new technologies and global digital networks, including the Internet, have so far helped us to move mostly toward "collective stupidity" (H. G. Wells's phrase), rather than toward global intelligence. There is no reason to believe that our machines will do better when left to their own devices, any more than our offspring have done, when left to theirs. Again, our ability to redirect the current evolutionary course will largely depend on our ability to change our mode of thinking and behavior not only in relation to our technology, but also in relation to each other and to all other forms of life on earth.

The present book is a very modest first step in the long and arduous journey toward global intelligence. Before proposing some concrete ways in which we can begin moving toward this goal, I attempt to clear the path by diagnosing the current ailments of our worlds of education and learning. I suggest that these worlds are largely dominated by a disciplinary mentality that organizes our cognitive and learning activities, as well as all other human transactions, indeed reality itself, in terms of linear power relations, engaged in a continuous struggle for achieving and/or preserving hegemony. Perfect examples of a disciplinary mentality are not only many of the ancient and not-so-ancient schools of religious and/or spiritual wisdom, but also their sworn enemy and symmetrical opposite, mainstream contemporary science, which prides itself on being largely secular, materialist, and democratic.

According to the Western, mainstream scientific paradigm, knowledge is objective, abstract, and universal, reflecting immutable physical laws. It involves a highly regulated and structured protocol of experiment and critical (dis-)proof, which continually approximates, although it may never reach, universal and eternal truth, valid in all places and at all times. Those claiming to be in possession of such truth, or of parts of it, or at the very least of the paths and key to it, are veritable gatekeepers who will share their knowledge or "let others in" only for a price. What I have just described, in an oversimplified form, is the famous equation of knowledge and power that, despite numerous critiques and challenges from various philosophical and scientific quarters, has long been an explicit or implicit model of scientific thinking and practice in academic and nonacademic circles, especially in the United States and Western Europe.

Whereas such disciplinary models of knowledge are deeply rooted in the European philosophical (and occult) tradition, they find their equivalents in the philosophical or religious traditions of other cultures as well, which explains why they have been exported so successfully, in the past century or so, practically all over the world. They remain operative, in many different forms and variations—be they scientific, technological, ideological, political, or religious—not only in large sections of academia, but also in many other fields of human endeavor throughout the planet.

In the present global circumstance, we need to transcend precisely this equation of knowledge and power that underlies the disciplinary paradigm. One might argue that various mentalities of power have served well certain cultures or, rather, restricted groups within those cultures, for short historical terms. In the long run, however, they have placed a severe strain on humanity and its habitat as a whole, threatening to impede further human development, if not to arrest it altogether. Given the experiences of the last few millennia, to say nothing of the past century—undoubtedly one of the worst in the known history of humankind in terms of destructive excesses—it is high time to abandon these mentalities and look for other ways of organizing human relations, as well as our cognitive and learning processes.

An approach to knowledge and learning oriented toward global intelligence will require philosophical and scientific presuppositions entirely different from those of a disciplinary mentality. It does not presuppose that knowledge is power, but only that power produces certain forms of knowledge, which may become irrelevant or transfigured in other, non-power-oriented, reference frames. It involves not only remapping traditional knowledge, as it is acquired, accumulated, and transmitted by various academic disciplines, be they scientific or humanistic, but also generating new kinds of knowledge from an intercultural perspective.

In turn, such a perspective presupposes a mutually enriching interplay of what various cultures perceive as universal or global and what they perceive as particular or local. Although these perceptions may vary widely from culture to culture, many cultures explicitly or implicitly employ this kind of distinction, whether in a positive or a negative manner. If some cultures, or at least some of their members, appear to perceive the global and the local as mutually exclusive or antagonistic, they do so because they basically operate within the reference frame of a disciplinary mentality. But, once this reference frame is left behind, the two notions become complementary and mutually supporting, rather than antagonistic and reciprocally destructive.

The mutually nourishing interplay of globality and locality finds its academic equivalent in the interplay of transdisciplinary or integrative knowledge and specialized or compartmentalized knowledge. It presupposes a holistic mode of thinking that looks beyond constituted academic fields and their divisions of knowledge, although it does not deny their usefulness. Yet, a global mode of thinking takes into consideration the fact that knowledge is "local" not only in terms of proper limits or confines (which should, however, remain flexible and open to redefinition), but also in terms of its historicity. Knowledge is always bound to a specific time and place, to a specific culture or system of values and beliefs or, indeed, to a specific lifestyle. A global approach attempts to identify the cultural specificities of knowledge, explore commonalities and differences among them, and negotiate, if need be, among such specificities. It also presupposes that, in the process of exploration of cultural commonalities and differences in the way in which we acquire and utilize knowledge, new kinds of cross-cultural knowledge emerge through intercultural research, dialogue, and cooperation, and new kinds of integrative cognitive and learning processes become possible.

A global approach to knowledge, moreover, implies the recognition (present, for example, in contemporary Western general systems theory, but also in early Buddhist thinking in India, early Taoist thinking in China and, later on, in Islamic Sufism) that there are many levels, or reference frames, of reality with their own specific "logic" and operating principles. In turn, these levels may be interconnected either in a chainlike fashion—according to what one calls "linear causality" in philosophy and science—or in a looplike, interdependent fashion, mutually nourishing each other, that is, according to mutual or reciprocal causality. As we move from disciplinary to interdisciplinary and then to transdisciplinary reference frames, as well as from monocultural to intercultural to transcultural or global ones, new levels of reality emerge, as well as new kinds of knowledge. At its broadest theoretical level, a cognitive and learning approach based on global intelligence concerns itself with and explores such questions as: What are the conditions of the possibility of the emergence of transcultural and/or transdisciplinary knowledge? How does such knowledge differ from, but also involve, cultural or local knowledge? How can it be communicated or taught? What uses can this kind of knowledge be put to and whom does it serve? What organizational and institutional forms might it take?

Whereas the present study does not seek conclusive answers to these questions, because, in any case, such answers can emerge only through extensive, intercultural and transdisciplinary research, dialogue, and co-operation, it does attempt to articulate certain guiding principles that could help us steer human (self-)development toward global intelligence. For this purpose, I have divided the book into three parts, corresponding to three main areas of knowledge that I consider essential for the development of a genuinely transdisciplinary field of intercultural studies: (1) global studies, currently conducted mostly within the social sciences, such as international relations, sociology, political science, and political economy; (2) environmental studies, conducted mostly within the life sciences; and (3) global education, addressed mostly by the academic field of "higher education," but also increasingly by a number of non-academic, national and international, governmental and nongovernmental organizations, concerned with developing new ways of learning and research in the context of globalization.

Another area of knowledge that is no less important for a trans-disciplinary field of intercultural studies includes the humanities and the performing arts, which largely constitute the object of *Remapping Knowledge*, mentioned above. Although the four main areas of knowledge in question are closely interrelated, academic and nonacademic researchers and practitioners within these areas are seldom aware of the issues that concern their colleagues or of the fact that many of them share the same concerns, especially at the theoretical and ethical levels. Consequently, they seldom talk to each other, let alone cooperate, across disciplinary and cognitive boundaries. My two companion volumes attempt, at least in part, to compensate for this regrettable lack, and can best be read in conjunction with each other.

In part I of the present study, I show that many of the most influential Western theories of globalization within the social sciences implicitly present themselves as being universal, rather than local, approaches. This kind of theoretical universalism can, for obvious reasons, be highly counterproductive in an intercultural environment. In order to avoid it as much as possible, I propose a distinction between globality and globalism. I define globality as an infinitely layered network of variously interconnected and interactive actual and possible worlds or localities. The human aspiration toward globality involves a desire toward (self-) transcendence that expresses itself as ceaseless world making and self-fashioning. In turn, I define globalism as the proper or improper expression of the aspiration toward globality. I suggest that most of the improper types of globalism that have so far manifested themselves in the history of humankind belong to various mentalities of power that aspire, and compete among themselves, for mastery over the planet.

In the light of the distinction between globality and globalism, I suggest that we rethink the main issues related to globalization, such as spacetime compression, the interrelationship between the global and the local, order and disorder, cultural identity and difference, and so on. It is at this point that I bring into my argument what I would like to call the *emergent ethics of global intelligence*. In my view, this ethics ought to be based on a mentality of peace, defined not in opposition to war, but as an alternative mode of being and acting in the world, with its own system of values and beliefs, and reference frame.[4] I call it "emergent" because it depends, no less than global intelligence, on continuous intercultural dialogue, research, and cooperation.

It is from such an ethical perspective that one may judge certain kinds of globalism (and localism) as proper or improper manifestations of the global aspiration. And it is also from this ethical standpoint that I propose to develop the principles and methodologies that ought to inform a reconstructed, transdisciplinary field of intercultural studies. This field will hopefully be able to transcend the current ideological and political impasse of cultural studies, especially as practiced in the West, while preserving and reorienting some of its valuable insights. It may also become an important vehicle for creating global learning environments that will nurture continuous human self-development.

The main principles and methodologies of transdisciplinary, intercultural studies that I briefly explore at the end of part I and develop further in *Remapping Knowledge* include: intercultural resonance (rather than mimesis); dialogical or responsive understanding; intercultural comparative analysis; intercultural and global self-awareness (rather than cultural critique or critical thinking); nonviolent contact and learning experiences at the liminal intersections of various cultures; and intercultural communication and cooperation, based on a mentality of peace. I also suggest a number of learning and research models that should initiate a much needed, comprehensive study of and dialogue among world cultures, not only from a local, national, or regional perspective, but also from a global one. The implementation of these models would involve a sustained, collective, and cooperative effort on the part of learners, educators, scholars, researchers, and other practitioners from academic and nonacademic fields throughout the world.

Part II takes up the main concepts developed in part I and places them within the framework of natural science, where they are equally operative, at both the theoretical and the practical level. I focus especially on the life sciences (with a very brief detour through elementary particle physics), because I consider them as being among the most important factors in the current drift toward globalization. Of course, Western-style science as a whole has now become a global subculture in its own right, and its (positive or negative) planetary impact can hardly be ignored.

Whether the worldwide reforms I envisage here will take place or not will greatly depend on the kind of science and corresponding technology that humankind will collectively develop in the next few decades. As it has by now become clear, I do not believe that the mainstream scientific approaches and practices that are currently in force are conducive to global intelligence. On the contrary, they may seriously impede human development. In order to make my point, I examine some of the main theoretical assumptions behind one of the most powerful trends in Western-style, mainstream science, namely reductionism, showing its negative ethical implications and practical consequences in a global environment.

Specifically, I look at the objectivist and universalist claims of the sociobiological version of Darwinian evolutionary theory in the contemporary life sciences and of the "theory of everything" in elementary particle physics. To me, these theories constitute good examples of the temptations of a totalitarian kind of interdisciplinarity and interculturalism, associated with a reductive, disciplinary mentality. The more a certain scientific or interpretative method claims to be interdisciplinary or cross-cultural, the more it claims to speak on behalf of all disciplines and cultures and to produce a unified picture of the world that supposedly applies in all places and at all times.

On the positive side, I look at scientific models outside reductionism that may be fruitfully adopted and developed in a global context. For example, I discuss in some detail the nonlinear models developed by general systems theory and applied in chaos and complexity theory, as well as in the environmental sciences in the West. But these nonlinear models are equally present in early Buddhist thought in India, early Taoist thought in China and, later on, in the Sufi Islamic tradition, thus requiring an extensive comparative analysis that I can only briefly develop in the present book. Such nonlinear models might be better suited to global, intercultural learning and research environments, not only because of their intercultural dimensions, but also because they show the basic interdependence of all physical and human events and do not reify local theories and practices into "objective," "irreversible" realities. On the contrary, they reveal the great freedom, but also the great responsibility that each human being has in creating or reversing a certain state of affairs in our world.

I then propose an ecological model of science, grounded in a mentality of peace that has constantly been present in many cultural traditions throughout known human history, even though it has often been misinterpreted, obscured, or pushed into the background. Here I mainly focus on its principles and practices as developed by early Buddhist, Taoist, and Sufi thinkers, as well as by contemporary, Western-style deep ecologists. I argue that from the standpoint of this irenic, ecological model, there is ultimately only one kind of science, which is human science, comprising any number of branches of learning and knowledge production, including the physical sciences, social sciences, and the humanities and the arts.

I conclude, however, that the deciding factor in human development is not so much how effective a scientific theory might be, or how close it might be to the truth, or how easily it can be turned into a "powerful" technology. Rather, it is our daily ethos or practice. We might possess the "best" scientific theory, such as deep ecology, the "best" information and communication technology, such as digital and quantum computers, or the "best" political system, such as democracy. And yet, we may end up putting them all to inappropriate and counterproductive uses, as we have seen happen repeatedly throughout the history of scientific and technological development, as well as the history of world politics.

Part III explores the conditions of the possibility of global learning and of local–global learning environments, oriented toward global intelligence. I focus especially on our institutions of higher education that train our scientists and other members of the world elites entrusted with directing human affairs. These institutions could thus play a decisive role in developing learning principles and practices that will reorient such world elites toward global intelligence. In the first chapter, I offer a brief history and a general assessment of the current assumptions and practices of Western academia, especially in North America and certain parts of Europe. Then I propose the development and implementation of a university model designed to foster the kind of local–global learning environments and intercultural, intellectual climate that are needed for sustainable human development in the next few decades.

In the second and last chapter, as well as in the appendix, I outline concrete proposals for institutional reforms and educational and research programs, designed to help create local–global learning environments throughout the planet. Chiefly among them is an academic blueprint for an advanced, doctoral program in Intercultural Knowledge Management (IKM) that would train local–global elites in the spirit of global intelligence and sustainable human development. The transdisciplinary and intercultural nature of the IKM research projects and academic curricula require that students move between institutions in several regions of the world, as well as across departmental divides at any single institution. A global perspective will give them the intercultural responsive understanding needed to bring together specialists or experts from various fields and from several cultures in order to design and execute transdisciplinary and intercultural projects that none of these experts would know how to implement on their own.

From the first year of their studies, IKM students begin to acquire and combine theoretical and practical knowledge in order to address real-time, local and global issues. They organize their curricula and research programs around the concrete problems they are asked to solve. They form cross-disciplinary and intercultural teams to work on viable solutions to specific, real-world problems, rather than through the codified practice of a particular academic discipline or culture. IKM students will thus build the capacity to identify and address potential socioeconomic and other types of problems before they develop into crises that threaten the peaceful development of world communities or diminish the diversity of world resources. They will also be called on to design workable, realistic blueprints for the sociocultural and human development of their countries or regions, based on the best traditions of wisdom available in their cultures, as well as in those of others, and on the most cherished aspirations and ideals of their people.

Although the present study concentrates mostly on Western-style, higher education and advanced scientific research, the guiding principles that I suggest should work reasonably well at other educational levels and in other parts of the globe. Whereas I largely focus on the cultural worlds with which I am most familiar (the United States and Europe), I attempt to take into consideration other worlds as well. Indeed, I emphasize again and again that any of the conclusions reached and principles advanced here are conditional and dependent on extensive, intercultural dialogue, research, and negotiation that will undoubtedly modify and enrich them in unexpected and delightful ways.

I would therefore like to conclude this introduction with an appeal to teachers, scholars, researchers, educators, and other practitioners concerned with the future of our planet, to start engaging in such worldwide, cooperative, and peaceful interactions, based on the present or similar proposals. It is my fervent belief that it is only in and through dialogic, intercultural learning processes, and not through obsolete, agonistic models of human interaction—such as the so-called wars on poverty, disease, drugs and, most recently, terror—that global intelligence will eventually begin to emerge and that further human development will become possible.

I

Cultural Theories and Practices of Globalization in Social Science

1

Globalization and Contemporary Social Science: An Intercultural Approach

A large number of the current cultural approaches to globalization in the social sciences are based on a Western notion of modernity as a construction of capitalist, middle-class mentality. This mentality presumably originated in the European Renaissance with the rise of a Protestant work ethic, was consolidated during the Enlightenment with the ideas of scientific rationality, order, and progress, and has now spread to many parts of the globe. In contemporary social studies, the Western theorists that continue to promote the capitalist mentality in a global framework are known as "neoliberals." The neoliberal position is well articulated, for example, in the influential work of Herman Kahn and his research group at the Hudson Institute, in such volumes as *The Next 200 Years: A Scenario for America and the World* (1976), *The Coming Boom: Economic, Political, and Social* (1982), and *The Resourceful Earth: A Response to Global 2000* (1984). This position was developed in response to a report on the *Limits to Growth* (Meadows et al. 1972, 1993), associated with the Club of Rome (a distinguished international group of scientists, businesmen, and policymakers, especially influential in the late 1960s and the 1970s), which advocated global restraint in economic development and responsible ecological practices on the part of the multinational corporate business community.

In turn, the most prominent twentieth-century critics of the neoliberal position are mostly, but not exclusively, of Marxist and Weberian descent. Among the best-known early critics of modern, industrial capitalism, one may mention Karl Marx, Max Weber, Theodor Adorno, Max Horkheimer, Walter Benjamin, Karl Mannheim, Antonio Gramsci, and Raymond Williams; recent prominent critics of postmodern, network capitalism include Jürgen Habermas, Pierre Bourdieu, Fredric Jameson, Jeremy Rifkin, Terry Eagleton, and Anthony Giddens.

The Marxist and Weberian cultural approaches to modernity are countered by postmodernist theories of Nietzschean and Freudian inspiration that are equally critical of middle-class mentality, but mostly from the standpoint of a will to power conceived as an unfettered, perpetual play of forces that arbitrarily constructs and deconstructs all social and cultural hierarchies. The best-known twentieth-century representatives of these postmodernist approaches are Georges Bataille, Michel Foucault, Gilles Deleuze, and Felix Guattari, among others.

The neoliberal, post-Marxist, and posmodernist theoretical positions are equally present in contemporary global studies. There are also those sociologists and political scientists in the global field who attempt to go beyond any specific, narrowly defined, ideology, adopting mostly an eclectic, descriptive stance.[1] Indeed, a handful of them also recognize that there is much more to globalization than its economic, technological, and political dimensions and that its cultural dimensions are equally relevant. Even more significantly, they argue that these cultural dimensions should no longer be reduced to the universal proliferation of the Western-manufactured concepts of modernity and postmodernity and their attending consequences.

For example, Roland Robertson aptly notes that most Western social scientists concentrate almost exclusively on the global economy, as well as, one might add, on geopolitics and information and communication technology. This exclusive emphasis "exacerbates the tendency to think that we can only conceive of global culture along the axis of Western hegemony and non-Western cultural resistance" (Robertson 1992, p. 113). The processes of globalization assume a wide range of national and regional forms that simply cannot be reduced to the perspectives generated by Western modernity without creating gross misperceptions of such processes (Featherstone 1995). So any viable theoretical approach to globalization should also be able to account for its enormously diverse cultural manifestations.

An important global cultural trend is the environmental movement, or at least that strand of it which steers free of fundamentalism on both the right and the left ends of the political spectrum. This nonviolent, peaceful ecological trend, also known as "deep ecology" (Naess 1973, 1989), generally avoids cooptation by the neoliberal forces of global capitalism. At the same time, it attempts to stay away from the antiglobalization forces that believe in and employ combative, activist strategies and tactics, ranging from political militancy all the way to armed resistance and terrorism. It goes well beyond conventional efforts to preserve the natural environment and seeks to transcend ideological and political factionalism by implementing a holistic approach to human development and biodiversity on a global scale.

From the standpoint of an emergent ethics of global intelligence, the nonviolent ecological trend seems to be the most promising and needs further theoretical and practical development, especially in an intercultural context, so that it can become a worldwide, mainstream way of thinking and acting. I shall discuss it at some length in part II, although I shall refer to it occasionally in this chapter as well. In what follows, I shall mostly reevaluate some of the mainstream cultural–theoretical approaches to globalization, drawing on the work of the more prominent social scientists in the field, whether they belong to the post-Marxist, postmodernist, neoliberal, ecological, or the eclectic camp.

At the same time, I shall propose an alternative approach that does not exclude the other approaches, but reinscribes them in an intercultural reference frame, redirecting them toward global intelligence. To this end, I would first like to sketch a tentative theoretical model that can provide a plausible world historical account of globalization and will also allow for various forms of ethical intervention in the process itself. Starting from this model, we can then rethink all of the major sociocultural issues associated with this phenomenon.

1 Globalism, Globality, and the Local–Global

To begin with, it would be useful to elaborate on the distinction between globality and globalism that I mentioned in the introduction. As we have seen, globality may be described as an infinitely layered network of variously interconnected and interactive actual and possible (or imagined) worlds or localities. Viewed in this perspective, our globe itself may be only a local manifestation of an even more complex globality. The organizing pattern of this larger reference frame may not necessarily unfold all at once, or become readily apparent to the political, scientific, artistic, or religious mind. Indeed, it might perpetually escape human understanding in both its individual and its collective guise. For this very reason, on the human level, globality seems to involve an aspiration toward self-transcendence. In turn, this aspiration, observable in most, if not all, human societies, expresses itself as ceaseless self-fashioning and world making.

Globalism, on the other hand, can be viewed as a proper or improper expression of the aspiration toward globality.[2] There are many kinds of globalism, some proper, some improper, some Western, some not. Most of the improper types of globalism that have so far manifested themselves in the history of humankind belong to a plurality of power-oriented mentalities that aspire, and compete among themselves, to hold sway over the entire planet. One may coin the phrase *globalitarianism* (on the model of totalitarianism) to describe such improper types. Familiar examples range from the Persian, Hellenistic, Roman, Napoleonic, and other empires to the various forms of socialist and communist internationals to some of the capitalist, consumerist global projects of today. Most recently, the globalitarian impulse has also resurfaced in the current U.S. administration's foreign policy of exporting U.S.-style democracy by military force to other parts of the world.

But there have also been proper forms of globalism that have, for example, manifested themselves since the Axial Age—that is, the age of the rise of the world's great religions and metaphysical systems in antiquity, as described by German philosopher Karl Jaspers (Robertson 1992, p. 101). That so far these forms have almost always had religion and faith as their point of departure does not mean that they belong to the religious impulse: on the contrary, it means that religion itself is a manifestation of the global aspiration that may in turn assume proper or improper forms.

In the light of the distinction between globality and globalism, one may rethink the notion of the global and the local and their interrelationship, which is at the core of all current cultural approaches to globalization. Most of these approaches assume, as mine does, that the global and the local are inextricably bound. Some posit a Hegelian or a Marxian dialectic between the two terms (Cvetkovich and Kellner 1997; Jameson and Miyoshi 1998). Other social thinkers approach the question from the standpoint of the Western philosophical issue of universalism versus particularism, which they also attempt to solve dialectically (Wallerstein 1991; Robertson 1992).[3] It is not implausible to assume that the local and the global are interdependent, or that their interrelationship can be described in terms of a dialectics of the universal and the particular. For example, Robertson's thesis of universalism and particularism as inherent in the global circumstance has some analytic cogency, especially in its experiential aspects, involving the *expectation* of universality and particularity in global human interactions, rather than their actual manifestation. One remains skeptical, however, precisely about the *actual* manifestations of the universal and the particular within the global field.

The problem, therefore, does not arise so much from the fact that Western social scientists view the relationship between the global and the local as if it were another manifestation of the dialectic between the universal and the particular. Rather, it arises from the essentialist definitions of these terms. Western theorists of globalization tacitly assume that such definitions are pre-given or universally accepted. Yet no two Westerners, let alone members of other cultures, might agree on their meaning and content, or on which phenomena might be defined as global and which as local. These definitions would, therefore, become effective within an intercultural framework only if they emerged through extensive intercultural research, dialogue, and semantic consensus.

Moreover, the various forms of dialectics themselves, not to mention binary thinking, are local, rather than universal, logical categories in the first place. Even the Japanese term "glocal," adopted by Western-style global studies, will always be defined in a local manner, or according to a globally dominant local view. Thus, dialectics (whether Aristotelian, Hegelian, Marxian, or any other kind) might mislead more than help us within a global reference frame.

We should not, then, automatically embrace a dialectics of the global and the local or automatically link these terms to a universally homogeneous, philosophical notion of universality and particularity. Instead, we should acknowledge that there is no overarching, cross-cultural, concept of globality available to us at this time, but a multitude of them. In other words, we should recognize that all we have are local theories of the global (King 1997; Featherstone 1995).

Contemporary ecological thinking equally recognizes this fact, for example, in the paradoxical formulation of American essayist and farmer

Wendell Berry: "Properly speaking global thinking is not possible" (Hawken 1993, p. 201). Here Berry has in mind an earlier ecological mantra: "Think globally, act locally." This could in turn be reformulated, more in line with his views, as "Think locally, act globally." Yet, if our local theories are to become genuinely useful as a basis for global action, they need to undergo a thorough and extensive comparative and historical analysis by intercultural and transdisciplinary teams of researchers and practitioners, engaged in worldwide dialogue and cooperation.

But we also need to reconsider our ideas of the local within an intercultural reference frame. In this regard, it would be useful to draw a distinction between localism and locality. Localism is a mirror image of globalism and can often take improper forms such as ultranationalism, ethnic intolerance, racism, religious and ecological fundamentalism, and so forth. Thus, recent examples of improper localism are not limited to the intolerant, close-minded attitudes of some monocultural or totalitarian nation-states. Unfortunately, they also include the paranoid, ultranationalist, and racist attitudes that certain Western governments have displayed toward immigrants from Islamic and other so-called Third World countries, as well as toward their own dissenting citizenry, in the wake of the September 11 and other terrorist attacks against Western (utilitarian) global interests. Locality, on the other hand, can be defined as the specific geocultural and ecological space in which individual and collective human activities, contacts, and interactions take place. Locality, just like globality, involves multilayered human perspectives and experiences, ranging from those of the tiniest community or organism to those of planet Earth or multiplanetary systems.

Keeping all of the above distinctions in mind, I would like to propose the working concept of the "local–global" as a provisional way of understanding globalization, despite its somewhat awkward, hyphenated form. A local–global theory would be a globally oriented local theory, just as a local–global community would be a globally oriented local community or eco-cultural environment. Communities around the world do not engage in interaction with a universal or a global community—although they may often imagine such communities—but with other local communities, some of which may happen, at one historical juncture or another, to become globally dominant or visible as far as their ideas and ways of life are concerned.

The global visibility of certain communities at various times in human history does not, however, oblige other communities to accept their ideas, values, beliefs, and mode of behavior as the global or the universal par excellence. On the contrary, these communities will often perceive such ideas and ways of life as improper forms of globalism (e.g., imperialism), particularly if they are forced on them. And it is often this kind of globally dominant particular that is raised to the level of globality *tout court* and is made to engage in the various dialectics of the global and the local advanced by most of the current Western cultural approaches to globalization. By contrast, a globally oriented local community (no less than a globally oriented local theory) would be constantly aware of its interconnections with other, close or distant, communities (theories) and would remain open to free and mutually enriching interaction with all of these communities (theories).

Since all of our local theories do, however, posit an interdependence of the local and the global, I would further like to suggest that this interdependence be seen not in dialectical terms, but as a manifestation of mutual causality or causal reciprocity. This notion is all the more appropriate in the present context, because it has a long cross-cultural history. As we shall see in part II below, it can be found, for instance, in contemporary Western general systems theory, as well as in Buddhist, Taoist, and Sufi thinking. From the perspective of mutual causality, there is a form of interdependence between the local and the global that goes well beyond the linear and binary notions of dialectics. Each intervention at the local level reverberates or resonates throughout the global reference frame, changing its configuration in unexpected and not always ascertainable ways. At the same time, the global is not a summation, nor a subsumption, of all its local instances, but a continuously shifting reference frame that opens up, beyond any and all of these instances, toward larger reference frames. In other words, what appears as global in one frame of reference may appear as local in another, and vice versa.

The notion of local–global causal reciprocity should also allow us to go beyond what Roland Robertson identifies as "worldism," since it regards globality, in a nonessentialist and nonlinear way, as an emergent phenomenon, rather than as a linear, invariable, or constant entity. At the same time, this notion should not be construed as a form of relativism or pluralism, nor should it be seen as a new form of pragmatism or localism. We should be able to turn away from the conflictive, binary opposition of "worldism" and "relativism" altogether, by introducing a local–global ethical perspective, or mode of thought and behavior through which disputed questions in the global field are settled neither by benign tolerance nor by active neglect, neither by fiat nor by force. Rather, they are resolved by continuous dialogue and peaceful interaction.

Furthermore, it is from this local–global ethical perspective that one may judge certain kinds of globalism (and localism) as proper or improper manifestations of the global aspiration. In the introduction, I have called this perspective *an emergent ethics of global intelligence.* Such ethics is in no sense pre-given, but can emerge only through intercultural dialogue and cooperation. This should not, however, prevent any of us, individually or collectively, from attempting to explore and define, at least provisionally, some of the conditions of the possibility of its emergence, as I have done throughout this study.

In view of the distinction between globality and globalism, as well as the nonlinear concept of the local–global, we may also wish to develop historical models of globalization within an intercultural, rather than a monocultural, reference frame. Robertson, for example, initiates this task in his examination of the deep cultural–historical roots of the current globalizing tendencies. He proposes a model of globalization that is based on the synchronic and diachronic interaction of four major sociological factors: societies, individuals, international relations, and humankind.[4] Although Robertson's historical approach is a step in the right direction, even a cursory look at his model reveals both its modernist and its limited, local, disciplinary perspective (based as it is on Western sociological concepts). For instance, he places the "germinal phase" of globalization in fifteenth-century Europe, even though one can trace globalizing tendencies in the recorded history of the world almost from its inception, as he himself is aware.[5]

Robertson also occasionally presents his outline as an evolutionary narrative. This might leave the false, and probably unintended, impression that globalization—together with liberal democracy—is an inevitable part of human evolution (a theory that is now fairly common in sociobiology and other natural and human sciences, as we shall see in chapter 3 below). But evolutionism is a relatively recent Western scientific theory that is not necessarily accepted as dogma in all parts of the world, let alone in the West, where creationist theories, for example, continue to challenge it vigorously.

By contrast, one might be able to build an intercultural, synchronic and diachronic, model of globalization with the aid of, say, Giambattista Vico's notion of *corsi* and *ricorsi* that he applies to historical events in general.[6] Thus, one may assume that globalizing tendencies of one sort or another are always present in human societies, but become dominant or visible at different times in different parts of the world. Their historical, cultural, and ideational content is never the same, but can widely vary according to the specific global circumstance, to use Robertson's apt phrase. At the same time, however, they are interconnected through complex feedback loops that reverberate throughout history, according to the principle of resonance that I shall discuss later on in this chapter.

Such globalizing tendencies become dominant, for example, during the times of the Babylonian, Hittite, Assyrian, Minoan, and Persian empires, during the reign of Alexander Macedon and the Hellenistic period, during the time of the Roman Empire, during the great migrations from Asia into eastern, central, and western Europe at the beginning of the Middle Ages (if not throughout the recorded history of the world), during the Arab and Ottoman conquests of North Africa, Middle East, central Asia, and central and eastern Europe, and so forth. One may also view such tendencies as forms of globalitarianism, or misuses of the global aspiration. In this respect, Arjun Appadurai, among others, points out that the "two main forces for sustained cultural interaction have been warfare (and the large-scale political systems sometimes generated by it) and religions of conversion, which have sometimes, as in the case of Islam, taken warfare as one of the legitimate instruments of their expansion" (Appadurai 1996, p. 27).

One may mention Christianism (to be distinguished from Christianity) as another such example of an improper, globalitarian drive—witness the various crusades in the Middle Ages and the frequently forced Christian conversion of the conquered colonies in later centuries. We should also distinguish radical Islamism from Islam, thus taking exception to Appadurai's invidious mention of Islam in the preceding citation. In this respect, the current proselytizing efforts of various Christian churches in eastern Europe and Russia, or the various Hindu or Islamic religious transnational networks, when disengaged from identity politics and capitalist consumerism, are not necessarily improper forms of globalism and may contribute to a healthy intercultural dialogue.

One may additionally point out that just because a few contemporary fundamentalist forms of Islamic, Christian, or Hindu religion seem again to display globalitarian tendencies, this fact does not change the nature of these religions as proper manifestations of the global aspiration. One must also not forget that commerce, artisanship, and learning, for example, were—and still are—equally important, and not always necessarily improper, conduits of global cultural interactions throughout the history of humankind. Pilgrims, travelers, merchants, peddlers, missionaries, adventurers, scholars, teachers, scientists, physicians, students, astrologers, necromancers, artisans and artists (architects, master builders, blacksmiths, coppersmiths, jewelers, printers, weavers, horse tamers, painters, sculptors, poets, musicians, dancers, actors, circus performers, etc.), and even "guest workers" (skilled and unskilled migrant laborers, slaves, and other types of foreign workforce) were all part of the global cultures of the past.

We should, therefore, not be too hasty in proclaiming radical shifts in the history of humankind, as many contemporary Western-style, social scientists, including Appadurai, do in regard to the current globalizing trends. In the end, the question is again one of definition, of evaluating and determining, from the perspective of an emergent ethics of global intelligence, what counts as a new paradigm. This question effectively boils down to that of the mentality or the system of values and beliefs prevalent in a particular community or society, which in turn generates its specific mode of thought and behavior, as well as its specific historical course. There are at least two kinds of history that materialist historians often conflate: the history of the development of a certain mentality, and the history of change or shift in mentalities. The history of modernity and its technological exploits belongs to the former and can hardly be heralded as a rupture or discontinuity in human mentality. Massive migrations and cultural displacements were part of global movements before, but did not effectively change the mindsets that drove them in the first place.

On the other hand, there are certain periods in the history of humankind when shifts in mentality do become possible: for example, during the sixth and fifth century B.C. with Gautama Buddha, Pythagoras, the teachers of Tao, and others; during the first century A.D. with Jesus of Nazareth and his disciples; and during the twentieth century with Mahatma Ghandi, Martin Luther King, and other leaders and participants in global movements advocating social change through nonviolent means. The momentum of these movements is unlikely to be lost in the twenty-first century.

At present the global circumstance may seem, at one level, to be unfavorable to a radical change in mentality—witness, for instance, increases in violent conflict, civil disorder, lawlessness, environmental degradation, terrorism, nihilism, political cynicism and opportunism, and so on. At another level, however, peace-oriented mindsets seem to be reemerging throughout the world. There is an increasing global awareness that violence, the depletion of world resources, and human selfishness, greed, cruelty, and injustice have a devastating impact on the quality and diversity of life on the planet and are therefore unsustainable modes of human behavior. In this respect, one could look at the massive, worldwide manifestations for peace on the eve of the U.S. government's military invasion and occupation of Iraq, not in terms of the "same old European pacificism" or "defeatism," as the U.S. and other global media largely presented them, but in terms of an increasing realization on the part of the world's populations that wars are no longer a sustainable way of solving global problems, no matter how just they claim or appear to be.

In light of the concept of globality that I have proposed, one might, then, look at the various human mentalities that have manifested themselves throughout history as creating various local and global subcultures. Such subcultures have in turn become "mainstream" or, on the contrary, have become submerged at different historical junctures. At the same time, all of these subcultures have engaged in complex and subtle feedback loops, with some of them attempting to subordinate the others. But these subordination efforts can succeed only if the subcultures in question belong to compatible or incompatible frames of reference. If they belong to incommensurable reference frames, as the power-oriented mentality and the irenic mentality do, they do not engage in reciprocal or mutually causal feedback loops, but each generates its own kinds.[7]

Even so, power-oriented mentalities have incessantly and, in part, successfully tried to put certain irenic values to improper uses, as we shall see throughout the present study, for example, when analyzing current, mainstream, notions of peace, life, death, love, cooperation, symbiosis, democracy, evolution, and so forth. Therefore, one of the tasks of an emergent ethics of global intelligence is to extricate these values from their current, power-oriented reference frames and place them in their proper, irenic frames.

Whether the paradigmatic shift to a mentality of peace will occur or become mainstream any time soon is a matter of the collective choice of more and more human beings all over the world. Ethical interventions on a local–global scale by individuals and groups are both desirable and possible, although the nature and modus operandi of such interventions have yet to be spelled out. But, one way in which we can begin to intervene, as I have just pointed out, is to rethink and revaluate, in terms of the basic principles of an emergent ethics of global intelligence, a number of major issues that are currently raised by Western and other social and cultural studies of globalization. In the rest of this chapter, I shall focus on three such major issues: the sociocultural effects of technology-induced compression of time and space on a global scale, world order and disorder, and cultural identity and difference.

2 Global Subcultures: Space and Time as Sociocultural Phenomena

The question of spacetime compression arises in connection with the rapid development of ever more sophisticated, electronic forms of information and communication technology (ICT). For instance, the World Wide Web almost instantaneously relays communications between users located at opposite corners of the earth. As such, the Web appears to create a cybernetic space that is paradoxically devoid of spatial and temporal dimensions. Within cybernetic space, physical obstacles and temporal distances seem to disappear as if by magic. As a result, the Web has been seen, particularly by a number of neoliberal scholars and practitioners, as an effective promoter of democratic values, facilitating a free circulation of people, ideas, and material goods across national boundaries throughout the world.

The neoliberal, optimistic view is countered by the post-Marxist, pessimistic contention that the technological compression of temporal and spatial distances does not equalize the human condition, but further polarizes it. On the one hand, this compression creates the global elite of mobility who possess "an unprecedented freedom from physical obstacles and unheard of ability to move and act from a distance" (Bauman 2000, p. 18). On the other hand, it creates a new class of dispossessed—the immobile class who are deprived not only of any freedom of movement, but also of any security on their own home ground. It also creates a new "absentee landlord" phenomenon, whereby the mobility acquired by those with capital leads to "an unprecedented disconnection of power from obligations: duties towards employees, but also towards the younger and weaker, towards yet unborn generations and towards the self-reproduction of the living conditions of all" (ibid., p. 9).

Post-Marxist and other social critics equally challenge the neoliberal claim that ever faster, electronic ICT improves the lives of billions of people, by making available a huge volume of material and immaterial goods in record time to a record number of consumers. For instance, John F. Kavanaugh notes that whereas a few individuals have used the latest information technology to circulate large sums of money faster and more efficiently all around the globe, this technology makes little impact on the lives of the world poor. Globalization based on ICT has so far benefited only a very small elite, while marginalizing two-thirds of the world's population (Kavanaugh 1991; Kavanaugh and Puleo 1991).

Post-Marxist and other social criticism thus also exposes the vicious circle between increased socioeconomic mobility, allegedly produced by information technology, and increased consumerism. The ever more rapid flow of material goods demands a rapid rate of consumption and therefore creates and actively promotes a consumerist mentality. This mentality functions on the principle of obsessive-compulsive desire, even though on the surface it appears as an exercise of free will. On every visit to the market place, consumers might feel that they are "the judges, the critics, and the choosers," because they could, after all, reject any of the choices on display. But what they do not have is "the choice of not choosing between them" (Bauman 2000, p. 84). The market preselects consumers as consumers and takes away their freedom *not* to consume.

From the standpoint of an emergent ethics of global intelligence, post-Marxist critics are certainly helpful in exposing the devastating socio-cultural and environmental effects of technologically induced spacetime compression and consumerist way of life. Yet, there are serious methodological problems with their approach within a global reference frame. One of these problems is the essentialist and universalist nature of their arguments, which they in fact share with their neoliberal counterparts. For example, consumerism as a way of life, no less than the capitalist mentality that drives it, is far from prevailing all over the world, including large areas of the Western world, such as Eastern Europe, parts of Central and Latin America, and so forth.

There are billions of people who are too poor and/or have no access to a shopping mall, let alone teleshopping or electronic commerce, to exercise their choice of being or not being consumers. Therefore, they cannot be lumped together with those who are part of the magic circle of consumerism. The few very rich exterritorials equally escape it, if for entirely different reasons. To belong to that magic circle, one needs to have a steady income and financial credit (including a credit card), that is, to belong to the various layers of the middle class.

In turn, the middle class (defined in Marxist, Nietzschean, Weberian, or other Western terms), with its consumerist culture in particular, may be globally vocal and visible, but it constitutes a relatively small minority around the world. Within that minority, moreover, there are many individuals who feel just like Bauman, Kavanaugh, and other social critics about consumerism. They have both a social conscience and a desire to live meaningful and productive lives, not least by serving others. So, post-Marxist and other social criticism unwittingly universalizes certain local capitalist values that may not be shared by the majority of the world population, let alone by many "capitalists" themselves. Its approach dramatizes the danger of speaking in abstract sociological and statistical categories such as capitalism, class, race, gender, and so on. This kind of essentialist approach can rarely take into account individual human beings with their complex nature and motivations that often go well beyond their purely economic and political transactions.

In this connection, one should not ignore an entire "ecology of commerce" (Hawken 1993) that advocates "natural capitalism" (Lovins, Lovins, and Hawken 1999) and small-scale global economies, based on low consumption, recycling waste, equitable sharing, and the dignity of work. This alternative business movement attempts to accomplish something more than merely trenchant critiques of multinational, corporate capitalism. Economists, business people, ecologists, civic activists, and politicians, such as Sharif Abdullah (1999), Wendell Berry (2001), Elise Boulding (1988), Kenneth Boulding (1978, 1990), Severyn Bruyn (2000), Herman Daly (1996), Amitai Etzioni (1988), Mikhail Gorbachev (1991, and with Mlynar 2003), Paul Hawken (1993, 1999), Hazel Henderson (1978, 1999), Amory and L. Hunter Lovins (1999), Gunnar Myrdal (1957), E. F. Schumacher (1975), Amartya Sen (2000, 1987), and others not only point out the destructive, unsustainable aspects of corporate culture, but also offer economic solutions, based, in some cases, on their own successful, alternative business practices.

For example, Henderson (1978) describes local and global "counter-economies," emphasizing the effectiveness of cooperative movements, worker self-management experiments, small-scale alternative economic ventures, and new patterns of production, ownership, and consumption. In her most recent book, *Beyond Globalization: Shaping a Sustainable Global Economy* (1999), Henderson lists a large number of individuals and organizations from all over the world that are involved in creating precisely this kind of sustainable global economy, as an alternative to corporate culture. Indeed, one might say that they constitute an important global subculture that will become, one hopes, increasingly mainstream in the not-too-distant future.

Post-Marxist, linear descriptions of the negative effects of global capitalism also ignore the subtle feedback loops that the local and global engage in. In this respect, no transnational company can play the "absentee landlord" game for long without seriously damaging its profits, let alone its image, precisely because all localities are perceived as interdependent, so that the global and the local have practically become inseparable. It is for this reason that successful transnational companies such as Aventis, Monsanto, Coca-Cola Inc., AT&T, Microsoft, Oracle, Shell, and others mount well-publicized local community initiatives, such as educational and research grants, environmental policies, cultural sensitivity programs, and so forth, designed to build up an image (whether accurate or deceptive) of responsible global citizenship.

There are also transnational corporate business organizations, such as the World Business Council for Sustainable Development (with a membership of over 130 transnational companies from all over the world) that promote what they call "corporate social responsibility." According to Paul Hawken, in the United States alone, in 1992, there were at least 70,000 companies that "were already committed to some form of environmental commerce" (Hawken 1993, p. 13), and their number must by now be significantly higher. To these organizations one may add the large number of environmental and civic enterprises in Europe, particularly in Germany and Scandinavia, as well as in other ecologically conscious regions from around the world. Such efforts might admittedly not be sufficient or ecologically "deep" enough, not to speak of their mixed motivation, such as high-mindedness, profit, opportunism, or greed. Even so, they are bound to have a resonating and amplifying worldwide effect that might eventually contribute to a paradigmatic shift in our local and global business practices.

One would therefore be well advised, from a local–global perspective, to limit the validity of the spacetime compression effect, as described by some neoliberal and post-Marxist analysts, to the worlds or subcultures created through, rather than by, electronic ICTs. If, within a global frame of reference, one may speak of spacetime compression, one may also speak of spacetime dilation or expansion. It is true that, in one sense, our planet has now become a single, compressed locality, at least for the participants in certain global subcultures. In another sense, however, it has also become an infinitely complex place, where one can imagine or experience an endless number of intersecting and interacting spacetime configurations—from the microcosms and macrocosms of organic life to the multiple universes of contemporary astronomy and the parallel universes of the nonlinear physics and mathematics of complexity; from the mythical, circular, and seasonal time of traditional, small-scale communities to the cosmic time of intergalactic travel.

One should also take into account the manufactured or conventional nature of time and space, with different systems and units of measurement still competing in defining our notions of objective, or physical, or unidirectional time, as well as our notions of three-dimensional space; and with even more complex systems and units yet to be devised in order to measure the new multidimensional models of spacetime imagined by contemporary physics and mathematics. Nor should one ignore the traditional distinction between objective and subjective or interior spacetime that is still largely operative throughout the globe, not to mention the different ideas of mythical, religious, sacred, ritual-bound, imagined spacetime that shape the cultural spheres of a large variety of human communities. One may thus draw further distinctions not only between measured time and duration, but also between vertical or axial and horizontal time, as well as between the various nonlinear concepts of spacetime, to be distinguished from the linear concepts of space and time, viewed as separate, invariable entities.

In this multilayered perspective, the technological compression of spatiotemporal distances neither equalizes the human condition, as some advocates of neoliberal democracy contend, nor polarizes it, as their post-Marxist opponents would have us believe. Rather, it is a local–global effect operative within a circumscribed sphere of cultural reality, and cannot be treated as a universal phenomenon. As a local–global phenomenon, it may also be compared to other such phenomena, so that one can properly qualify not only its claims to universality but also its claims to uniqueness (witness Bauman's statement, already cited, to the effect that Western elites now possess an "unprecedented freedom from physical obstacles and unheard of ability to move and act from a distance").

One may argue, for instance, that there are other forms of spacetime compression, such as out-of-body experiences, telepathy, cosmic travel, and so forth, which are even more spectacular, from a local, Western standpoint, than the feats of Western elites and their fancy technologies. There are world subcultures in which shamans, yogis, lamas, and others can accomplish instant, long-distance communication and transportation, whether terrestrial or extraterrestrial, without the aid of electronic technology. They place themselves in an axial or inner–outer dimension in which time as unidirectional flow does not operate. In such mindsets, the individual is not a social agent, but a form of cosmic being–awareness for which three-dimensional space and unidirectional time look like rather primitive models of conceiving and operating with "reality," or locality. Of course, such phenomena have in turn largely been dismissed by Western modern science as paranormal, or as psychological illusionism or worse, and are only now beginning to be accepted very slowly and partially as legitimate objects of study, let alone as valid, "objective" facts.

My point is simply that what might appear as spacetime compression in one global subculture might not be necessarily experienced as such or in the same way within another subculture; and that different cultural notions and experiences may easily lead to misunderstanding and conflict. In sum, one should take into consideration the cultural dimension of any concept of space and time, whether they are seen as objective, universal givens or as subjective and relative, local factors. Cultural spacetime may vary greatly from civilization to civilization, indeed, even from one human community to the next. Therefore, it is essential for any theory of globalization (and globality) to explore and understand these variations, and for any proper globalizing effort to submit such concepts to the same process of sustained intercultural research, dialogue, and negotiation that is appropriate for other culture-specific notions and practices.

Indeed, as I have already pointed out above, within a global reference frame, various localities, which can also be defined as specific forms of the spacetime continuum, constitute various global subcultures that engage in complex, constructive or destructive, feedback loops. In turn, such subcultures can be seen as forming a plurality of possible or actual worlds that crisscross each other and that may be compatible, incompatible, or even incommensurable. In other words, they can be part of more than one local–global frame of reference, as I shall detail in chapter 4, when discussing the possibility of access to worlds that are organized on principles other than power. Global subcultures, therefore, are not necessarily limited to traditional cultural boundaries, but may extend across various world cultures. Industrial and network capitalism, Western-style science, democratic or autocratic forms of social organization, religious denominations, digital ICT, and so forth are all elements of such global subcultures that extend beyond the boundaries of a single world culture or civilization, let alone one country or nation-state.

From the emergent ethical perspective of global intelligence, one should also reconsider the assumption shared by most scholars and practitioners in global and cultural studies that globalization is largely a technology-driven change. This assumption often goes hand in hand with the belief that Western-style, digital technology is another inevitable step in the triumphant march of humanity toward universal happiness, based on material self-interest. Yet, there is nothing inevitable about digital technology, nor about the globalizing trends that it allegedly sets in motion on an irreversible course. One may argue, on the contrary, that a certain mentality or mindset creates a certain technology, which in turn promotes and amplifies this mentality, but without radically transforming it. As I have already noted in the introduction, radical historical shifts do not come from changes in technology, but from changes in mentality that will then generate other technologies. In other words, mentality and technology equally interact through constructive (or destructive) feedback loops, according to the nonlinear principle of mutual causality or causal reciprocity.

Nor should one proclaim the universal reach and unidirectional impact of Western-style ICT and new media, as many neoliberal, post-Marxist, and postmodernist cultural theorists do. For these theorists, a certain type of media-promoted, contemporary popular culture, with its material-consumption- and status-driven obsessions, often becomes the only proper, or even conceivable, domain of cultural activity, at least as far as postmodernity is concerned. Such an ahistorical approach is ironically at odds with traditional Marxist teaching. It ignores most of the world's long-standing cultural traditions or, even worse, dismisses or distorts these traditions in the name of modernity and progress.

An interrelated problem is the view of the human imagination as an exclusively mimetic faculty, a view that is also shared by neoliberal, post-Marxist, and postmodernist theorists alike. To give just one example, a sophisticated, postmodernist advocate of globalization such as Arjun Appadurai seems to believe that locals and migrants cannot exercise their imagination freely, but only within the preset framework of the new media and of other (popular) cultures. They can mimetically imagine only what is already available in other societies, rather than alternative ways of life that may equally stem from their own traditions (Appadurai 1996, pp. 5–6).

Yet, do all migrants search only for more plentiful food and clothing, more secure housing, and a better social status or group identity than they had at home, as Appadurai implies, or are they motivated by other values as well? Are they nothing but puppets and dupes of identity politics, or are they complex and creative human beings capable of further social, ethical, and spiritual development, beyond what is commonly available in their own and in other world cultures? In skirting these issues, Appadurai seems to lack precisely the kind of creative imagination that may catalyze genuine sociocultural change. Indeed, he never ventures beyond the network capitalism that is the object of post-Marxist critique. In fact, a post-Marxist theorist such as Zygmunt Bauman would probably classify Appadurai, however fairly or unfairly, as one of the postcolonial border crossers, an exterritorial, postmodernist member of the new global academic elite who help along, rather than resist, the unwanted consequences of a consumption-driven globalization.

But, despite their different, in some ways opposite, approaches to globalization, all post-Marxist and postmodernist cultural analysts give undisputed primacy to power and cannot imagine a world or a mentality outside it, any more than their neoliberal counterparts can. Specifically, they still subscribe to the rhetoric of victimization, guilt, blame, and resistance that many contemporary politicians, academics, and civic activists continue to employ ad nauseam. This victim-oriented strategy of power thrives on the mimetic mechanisms of envy and *ressentiment* and goes hand in hand with the mentality of greed that critics of postmodern capitalism deplore. For example, Bauman cites the findings of the United Nations' latest Human Development Report, according to which the total wealth of the top 358 global billionaires equals the combined incomes of 2.3 billion of the world's poorest people. Then he quotes Victor Keegan to the effect that "if the 358 decided to keep $5 million or so each, to tide themselves over, and give the rest away, they could virtually double the annual incomes of nearly half the people on Earth. And pigs would fly" (Bauman 2000, p. 70).

In turn, Appadurai employs the same kind of populist rhetoric when he refers to "the story of the American missionization and political rape of the Philippines, one result of which has been the creation of a nation of make-believe Americans, who tolerated for so long a leading lady who played the piano while the slums of Manila expanded and decayed" (Appadurai 1996, p. 30). Such statements and make-believe statistics smack of the demagoguery that both Bauman and Appadurai reject elsewhere in their studies. Is there more than a trivial connection between the piano playing of one dictator's wife and the decay of his country? Are there not countries in the world whose populations have fared as badly, or even worse, under dictators whose wives could barely write their own names, let alone play the piano? And even if the income of all the world populations would double or quadruple in the next few years, would that improve their lives, or the overall quality of life on earth? Would not most of that income sink back into the consumerist vicious circle that Bauman and other social critics indict?

Many post-Marxist and postmodernist cultural analysts seem to share with their neoliberal counterparts the unexamined, utilitarian assumption that income, that is, spending power, is what creates better human beings, in control of their destinies. The only difference is that whereas neoliberals believe in redistribution of material wealth through trickle-down economics and charity, their leftist counterparts believe in redistribution through heavy taxation and/or forced collectivization of private property. Both strategies have miserably failed in the past, and there is little chance that they will work any better in a global framework than they did in a local one. Should pigs suddenly fly and billionaires decide to part with their wealth, one would hope that they would spend their billions on less foolhardy projects. The radical question is not to redistribute the wealth of the world's populations—that communist program has irrevocably failed—but to imagine ways in which one can go beyond the notions of material wealth and poverty themselves as dominant marks of human worth.

To sum up my arguments so far, the emergent ethics of global intelligence should start from the recognition that global capitalism with its consumerist mentality and its linear notions of space, time, and human development is only a local–global subculture like many others. Within the globality of our planet, there may be—or one may imagine—many different worlds that are not primarily driven by the utilitarian, free-market logic described by Western-style, neoliberal, post-Marxist, and postmodernist theorists. Therefore, it is our task not only to identify or imagine such worlds, but also to work collectively toward their (re-)emergence as alternatives to the current ones, which have largely proven to be unsustainable.

3 Order and Disorder, Identity and Difference: A Local–Global Approach

An emergent ethics of global intelligence should also reconsider the current notions of global order and disorder, highly visible in neoliberal political slogans such as the New World Order, or in controversial book titles such as *The New World Disorder* (Jowitt 1993). These notions are often invoked in relation to the ambiguous role or "fate" of the nation–state in a globalized, political world. For instance, social and political analysts such as Kenneth Jowitt argue that, with the withering away of the nation–state and the disappearance of the bipolar division of the political world into a capitalist and a communist camp, no one seems now to be in control of world affairs, which have become unruly, indeterminate, and chaotic. This geopolitical chaos has cleared the way for the unrestrained, anarchic forces of the market to ravage the globe. Because the contemporary nation-state has largely lost its sovereignty and independence, effectively having no say in managing the national or global economies, its role is now reduced to administering affairs on behalf of multinational megacompanies and to policing local populations.

In turn, Appadurai (1996) also refers to the shifting role of the nation-state, but his view is generally optimistic. He hopes that a future, post-national, political order, created by various transnational civic networks will replace the current "disorder," assuring the protection of minorities and the equitable distribution of democratic rights in a new global society. At the same time, in typical postmodernist fashion, Appadurai regards "disorder" itself as an engine of global development. For instance, he conceives of globalization as chaotic, disjunctive flows of people, goods, money, images, and ideas. Invoking the (Western) scientific notions of order and disorder, he claims that chaos theory is the most appropriate perspective for viewing this phenomenon.

According to Appadurai, in a world of disjunctive global flows, the old scientific questions of causality, contingency, and prediction should be rephrased in terms of images of fluidity and uncertainty, which should in turn replace the older images of stability and order. He further claims that unless we develop a theory of global cultural systems that properly accounts for dynamic flow, we would be simply perpetuating our "illusion of order" (Appadurai 1996, p. 47). Here Appadurai obviously invokes the postmodernist (but also post-Marxist) critique of the Enlightenment's philosophical notions of rationality and order, dressed up in their modern, scientific garb.

The argument can equally be made, however, that if there is an illusion of order, then there is also an illusion of disorder: order and disorder are part of the same set of binary oppositions that can be viewed as cultural illusions or, neutrally, as operative cultural assumptions or practical fictions in the Vaihingerian sense.[8] In this respect, one may point out that images of chaos, uncertainty, and dynamic flow are at least as old as their symmetrical opposites. Examples range from the Homeric and Hesiodic cosmogonies and the Democritian and Lucretian atomic theories in the Western ancient world, to the sophisticated scientific concepts of order and disorder present in the Vedic tradition in India, or in the Taoist and Confucian tradition in China.

Appadurai's mimetic assumption, like that of other contemporary advocates of chaos, complexity, and emergence theory in the social sciences and the humanities, appears to be that for a seemingly chaotic historical moment, one needs to develop chaotic models. But the present time can also be seen, and has in fact been seen, as a reconfiguration of order within apparent disorder—witness the optimistic philosophical accounts of Francis Fukuyama (1992) and those of other neoliberal advocates of the free market, network capitalism, and the New World Order, based on militarily enforced, American-style democracy. In fact, as we shall see in part II below, the proponents of chaos and complexity theory in the physical sciences, in marked contrast to some of their colleagues in the social sciences and the humanities, are much more interested in order than in disorder. Specifically, they are interested in the ways in which "order" emerges out of "chaos," to paraphrase the title of Ilya Prigogine and Isabelle Stengers's well-known study (Prigogine and Stengers 1984).

Conversely, one can see all history, not just the present moment, as chaotic disorder. We can go back to almost any period recorded in human history and discuss it in terms of disjunctures between Appadurai's particular five kinds of flow or, for that matter, any other kinds. Again, the question is not of denying the cogency and explanatory usefulness (however extended or limited) of chaos and emergence theory as a local approach to globalization, but, rather, of becoming aware of its culture-specific nature. One would then avoid inflated claims about its universal applicability and would properly submit it to an extensive intercultural research and dialogue.

The same local–global approach should be adopted in considering various notions of identity and difference that are currently in vogue within the field of cultural and global studies. These notions are inextricably linked to the ideas of "culture" and "civilization" that are increasingly employed by contemporary students of globalization. Since in the present book I also frequently refer to "culture" (as well as to cultural and intercultural approaches), it might be helpful briefly to explore the current meanings and uses of that term, before I deploy my own uses of it, in chapter 2 below.

One may roughly discern two main concepts of culture in contemporary social science: the first one is an essentialist and substantialist view, tied, no less than the notion of globalism and localism, to a dialectics of the universal and the particular. In this view, culture is a durable, substantial and, ultimately, universal entity that determines the identity, coherence, and solidarity of a larger or smaller social group. In turn, cultural identity creates cultural differences, which are, as a rule, contingent, insubstantial, and nonessential and can eventually be resolved or reconciled in a universal culture.

The second view of culture is the symmetrical opposite of the first one: it raises cultural difference to an essential status and, consequently, sees cultural identity as a fluid, unstable, and insubstantial state in the ceaseless play of cultural differences. Above all, it perceives this play of differences as a conflictive or agonistic one. Postmodernist schools generally prefer the second view, whereas neoliberal, modernist, and other traditional cultural approaches, such as Marxism, prefer the first one. More often than not, the two concepts of culture equally engage in a contest for cultural authority and thus generate amplifying feedback loops, according to the principle of mutual causality.[9]

A good illustration of these conflictive concepts of culture can be found in *The Clash of Civilizations and the Remaking of World Order* by Samuel P. Huntington (1996), who moreover presses them into the service of a neoliberal vision of the New World Order, under the political and military hegemony of the United States. Huntington's book merits detailed, renewed scrutiny, if for no other reason than because it is reportedly intended as a possible foreign policy blueprint for the United States in the post–Cold War era and has the endorsement (at least on the cover of the paperback edition) of such influential American foreign policy makers as Henry Kissinger and Zbigniew Brzezinski.[10]

Huntington starts out by appropriately stressing the crucial role of culture in understanding the political and economic decisions and actions of various nation-states from around the world. He notes that people "define themselves in terms of ancestry, religion, language, history, values, customs, and institutions" (1996, p. 21). People also identify with cultural groups, such as "tribes, ethnic groups, religious communities, nations, and at the broadest level, civilizations" (ibid.). In turn, for Huntington, as for many cultural historians, a civilization is "the highest cultural grouping of people and the broadest level of cultural identity people have short of that which distinguishes humans from other species" (ibid., p. 42).

Acknowledging that his classifications are necessarily based on the kind of abstraction, reduction, and oversimplification that attends any paradigmatic way of thinking, Huntington then distinguishes among eight major contemporary civilizations: Sinic, Japanese, Hindu, Islamic, Orthodox, Western, Latin American, and possibly African (ibid., pp. 45–47). He adds that religion is a "central defining characteristic of civilizations" (ibid., p. 47) and notes that four out of the five great world religions, Christianity, Islam, Hinduism, and Confucianism are associated with his list of major civilizations. He also mentions Buddhism, but argues that there is no Buddhist civilization, and significantly omits Taoism altogether as part of Sinic and other civilizations.

One may object to the number of contemporary civilizations that Huntington distinguishes or to the ways in which he distinguishes them. But, overall, he does neither better nor worse in this respect than other traditional, Western historians of civilizations such as Oswald Spengler (1926), Arnold Toynbee (1934–1961), and Matthew Melko (1969), from whom he largely derives his general definitions. Much more objectionable, however, is Huntington's insertion of contemporary identity politics into his seemingly neutral descriptions. For example, he states from the outset that people "use politics not just to advance their interests but also to define their identity. We know who we are only when we know who we are not and often only when we know whom we are against" (Huntington 1996, p. 21). By the same token, civilizations are "the biggest we within which we feel culturally at home as distinguished from all the other thems out there" (ibid., p. 42).

Huntington then predicates that conflict will inevitably arise among various civilizations, because "it is human to hate" (ibid., p. 130). According to him, people, in order to define and motivate themselves, will always need enemies, such as business competitors, political opponents, and rivals in achievement. They "*naturally* distrust and see as threats those who are different and have the capability to harm them" (ibid., p. 130, my emphasis). In today's world, moreover, the "them" is more and more likely to be people from a different civilization. Huntington concludes that "cold peace, cold war, trade war, quasi war, uneasy peace, troubled relations, intense rivalry, competitive coexistence, arms races . . . are the most probable descriptions of relations between entities from different civilizations. Trust and friendship will be rare" (ibid., p. 207).

One of the most worrisome aspects of Huntington's book is his subordination of the great world religions to identity politics or, in my terms, his conflation of religion with religionism. This is most obvious, for example, in his treatment of Islam and what he calls "Islamic civilization." Playing the abstract game of statistics that Bauman and other contemporary sociologists also indulge in, Huntington contends that "Muslim bellicosity and violence are late-twentieth-century facts which neither Muslims nor non-Muslims can deny" (ibid., p. 258). He self-servingly cites James Payne's statistics in *Why Nations Arm* (1989). According to Payne and then Huntington, Muslims had a high propensity to resort to violence in international crises between 1928 and 1979, especially "high-intensity violence." For instance, they resorted to "full-scale war in 41 percent of the cases where violence was used" and engaged in "major clashes in another 38 percent of the cases" (Huntington 1996, p. 258).

Huntington's and Payne's statistics could be easily dismissed, if they did not involve fairly common Western, mainstream, assumptions and prejudices about other (Western and non-Western) cultures. Some of these assumptions and prejudices have arguably led to two devastating, largely Western-initiated, world wars in this century, possibly the bloodiest in all human history. One should also not forget that it was the government of a Western country, the "core nation state of Western civilization" to use Huntington's terminology, that dropped the first and only nuclear bombs on the population of another (non-Western) country. So it is at least disingenuous on the part of a Westerner to talk about "Islam's bloody borders" (ibid., p. 254), especially in view of the present global circumstance. Unfortunately, violence as a way of solving conflicts is a widespread global phenomenon and no single large culture or "civilization" is more—or less—prone than another to resort to it. Huntington indirectly undermines his own argument when he cites Payne's statistics according to which China's use of high-intensity violence in 76.9 percent of its crises far exceeded that of the Muslim states (ibid., p. 258).

Nor can violence employed by Muslim states, or any other state, be explained by such sociological factors as Islamic "demographic bulges" (great increases in Islamic youth population) or such "cultural" factors as the supposedly violent nature of Islamic religion. For instance, Huntington approvingly cites Payne's argument that Islam began as a religion of the sword and that it glorifies military virtues. Muhammad himself was a brave fighter and a skillful military commander, but "no one would say this about Christ or Buddha" (ibid., p. 263). Finally, according to Huntington, the Koran and other statements of Muslim beliefs contain few prohibitions on violence, and a concept of nonviolence is supposedly absent from Muslim doctrine and practice (ibid.).

To realize the invidiousness of Huntington's argument, one may simply remember Christ's statement that he comes with a sword to bring war, not peace, to his land, or the fact that Gautama Buddha belonged to the warrior rather than the priestly cast; or, conversely, that the traditional greeting among Muslims is "Peace be with you" (*Al-Salam Aleikum* or *Domtum fi Salam wa Amen*) or that Sufi teachings do not condone violence and conflict any more than their Buddhist, Taoist, or Christian counterparts do. For example, the prophet Muhammad says: "If a man gives up quarreling when he is in the wrong, a house will be built for him in Paradise. But if a man gives up a conflict even when he is in the right, a house will be built for him in the loftiest part of Paradise" (Frager and Fadiman 1997, p. 84). If anything, Huntington's and Payne's arguments highlight the ignorance of even well-trained Westerners about other cultures and religions (not to mention their own) and the urgent need for educating the world's youth about each other's—and their own—cultural traditions.

In fact, many of the world's religions seem to have arisen precisely as ways of containing or limiting violence, as a number of Western anthropologists and historians of religions have shown (Girard 1977, 1986, 1987). On the other hand, the origins of ethnic and nationalist violence can be traced back less to religious faith, as Huntington and other political analysts argue, than to the kind of mimetic rivalry that certain political elites from around the world engage in and fuel in their own populations. So it is not the case, as Huntington argues, that people use politics and religion to define their identity but, rather, that some politicians use people's need for faith and identity in order to achieve their own, narrow political ends. In mimetic political rivalries, the professed enemies—who have often been friends in the past, admiring and imitating each other's cultures—become mimetic doubles: they proclaim their radical differences, whether political or cultural, at the same time that they accuse each other of identical egregious behavior, and generally imitate each other's rhetoric and actions (Girard 1977).

Huntington himself unwittingly demonstrates this tendency when he argues, somewhat inconsistently, that it is not Islamic "fundamentalism" (i.e., religion) that is "the underlying problem" for the West. Rather, it is Islam, "a different civilization whose people are convinced of the superiority of their culture and are obsessed with the inferiority of their power." In turn, the problem for Islam is not the CIA or the Pentagon. Rather, it is the West, "a different civilization whose people are convinced of the universality of their culture and believe that their superior, if declining, power imposes on them the obligation to extend that culture throughout the world" (Huntington 1996, pp. 217–218).

For "people" in the foregoing citation, however, one would be well advised to read not the world's populations, as Huntington would have one believe, but certain political elites. So, it is neither Islam nor the West that are a problem for each other, but certain political leaders and their advisers. These include Huntington and a few of the Islamic authorities whom he cites—most of them educated at prestigious Western schools and thoroughly familiar with the kind of discourse that will catch the attention of Western political elites. It is not the "people" of various cultures, but members of these political elites that are obsessed with the superiority or inferiority of their power, set each other up as mimetic rivals, and then manipulate or force the rest of the population into following them.

Huntington and his kind of political adviser (a *Realpolitiker* of the Kissinger type) seem uncomfortable with the disappearance of the Great Schism, during which the world was divided between two superpowers (at least in their own minds), and the enemies were clearly defined political doubles. In the absence of such bipolar neatness, Huntington proliferates mimetic doubles by setting up several "civilizations," which he then postulates as engaging in the same conflictive relationship that was operative between the two superpowers during the Great Schism. According to this logic of power, masterfully described by George Orwell in *Nineteen Eighty-Four* (1949), when you have no enemy, you must create one. So Huntington lumps together Arabic, Persian, Turkic, and other cultures that originated in the Middle East, North Africa, and other parts of the Mediterranean basin, calls them an "Islamic civilization," and then pits them against "Western civilization," another huge conglomerate of widely diverse cultures. In turn, he arbitrarily divides the latter into a Catholic–Protestant and an Orthodox civilization, but calls only the former Western, this time using Christian, rather than Islamic religion, to create a mimetic conflict between the West and non-Western Russia and parts of eastern Europe. Thereby, he greatly oversimplifies the rich traditions of all of these cultures and religions, perpetuates some of the common misunderstandings about Islamic and Christian faith(s), and undermines his own professed goal of promoting genuine multiculturalism abroad.

Thus, although Huntington seems at first sight to give culture a pride of place in his argument, it becomes obvious that in effect he subordinates it to the will to power, just as much as other neoliberal, post-Marxist, and postmodernist political scientists do. For example, after he proclaims that a cultural approach is indispensable in understanding the complexities of global politics, economics, and finance, Huntington writes: "The distribution of cultures in the world reflects the distribution of power. Trade may or may not follow the flag, but culture almost always follows power. Throughout history the expansion of the power of a civilization has usually occurred simultaneously with the flowering of its culture and has almost always involved its using that power to extend its values, practices, and institutions to other societies. A universal civilization requires universal power" (Huntington 1996, p. 91).

Citing Joseph Nye's distinction (Nye 1990, 2004) between "hard power," based on one's economic and military strength, and "soft power," based on the appeal of one's own culture and ideology to others, Huntington subordinates the second type to the first: "What, however, makes culture and ideology attractive? They become attractive when they are seen as rooted in material success and influence. Soft power is power only when it rests on a foundation of hard power" (ibid., p. 92). Thus, as the hard economic and military power of a certain country increases, it produces "enhanced self-confidence, arrogance and belief in the superiority of one's own culture or soft power compared to those of other peoples." At the same time, this culture becomes increasingly attractive to other nations. Conversely, diminishing economic and military power produces "self-doubt, crises of identity, and efforts to find in other cultures the keys to economic, military, and political success" (ibid.). It is quite understandable, therefore, that as non-Western societies begin to enhance their economic, political, and military power, they "increasingly trumpet the virtues of their own values, institutions, and culture" (ibid.).

For Huntington, then, culture is ultimately nothing but soft power, or window dressing for hard (economic, political, and military) power. He deplores the declining power of the West and thus advances a double-binding argument. On the one hand, since the West is gradually losing its will to power, reflected in its (perceived) loss of hegemony over the entire world, it should also abandon its universalist cultural claims. Instead, it should attempt to adjust itself to the New World Order, where "younger" and more "aggressive" civilizations such as the Chinese and Islamic ones are beginning to assert their own hegemonic claims. (Observe the exquisite irony of calling such old civilizations as the Chinese and Islamic ones younger than the Western one.)

According to Huntington, the United States should, as the "core nation state" of Western civilization, encourage multiculturalism abroad and monoculturalism at home, rallying the Western and Central European states around it, as well as other states such as Canada and Australia, which belong to the same civilization, whether they like it or not.

The United States should therefore entrench and fortify its civilizational values at home, rather than attempt to stretch them too thinly abroad. On the other hand, if the West is not prepared to relinquish its hegemonic claims, then it should learn its lesson: "Culture, as we have argued, follows power. If non-Western societies are once again to be shaped by Western culture, it will happen only as a result of the expansion, deployment and impact of Western power. Imperialism is the necessary logical consequence of universalism" (ibid., p. 310). Needless to say, our current U.S. foreign policy makers have decided for this second course. Such a globalitarian course has never worked in the past and there is no reason to believe that it will work in the future. On the contrary, it threatens again to bring the whole world to the brink of a major disaster.

Huntington regards the American advocates of multiculturalism at home as committing cultural suicide and thereby further weakening and advancing the decline of the West. This is in line with his view that soft power is directly subordinated to hard power. But the opposite argument can also be made: it is the multiculturalists at home who unconsciously or deliberately further the hegemonic interests of the West by affirming its civilizational diversity or soft power and thereby defusing the virulent reactions against Western culture that Huntington perceives as coming from other civilizations. By paying lip service to diversity and placing in visible positions of power a few token individuals, carefully selected from the most politically vocal ethnic, racial, or gender groups, American multiculturalists in fact save the hegemonic values of their civilization. It is monoculturalists such as Huntington, the argument would conclude, that exacerbate the "clash of civilizations," by openly and crudely affirming their hegemonic goals.

If Huntington's history teaches us anything, it is that power has often fared best under various disguises, rather than through raw display, that is, that soft power can often be harder than hard power. This truth should be painfully obvious to those U.S. foreign policy makers who advocate preemptive strikes as a way of preventing terrorist and other military activities on the part of so-called rogue nations and political groups, inimical to the United States and its closest allies. Such displays of raw power have led, for example, to the current debacle in the Middle East.

By the same token, one can see the thesis of the decline of the West itself as yet another ploy of a will to power whose very nature is insecurity, paranoia, and an insatiable desire for continuous expansion and self-affirmation. For example, Nazi ideologists opportunistically used this Spenglerian thesis to justify Germany's expansionist policies before and during World War II. Elsewhere, I have described what I call the *ethopathology* of power (Spariosu 1997), of which Huntington's study is a perfect example. Without insisting on that analysis, here I shall simply note that my contention is not that the will to power is not alive and well, especially among certain political elites from around the world. It is the values of this global "master race" in the making that Huntington presents in essentialist and universalist terms, for example, when he says that it is "human nature" to hate and compete with others, to enjoy wielding power, and to resort to violence in order to achieve one's hegemonic goals.

Such a global subculture obviously exists and its mode of thought and behavior is exactly that described by Huntington (and Nietzsche). In this sense, his book is a valuable, if perhaps unintended, critique and caveat: it indirectly points to the disastrous consequences for humanity, should this global subculture based on a raw mentality of power continue to increase its worldwide influence. It also underscores the urgent need to educate young local–global elites in a different spirit, which might be one of the most effective ways of avoiding the kind of brave new world that Huntington, in the wake of Nietzsche, envisages for us.

2

Intercultural Studies: A Local–Global Perspective

An emergent ethics of global intelligence should certainly not ignore the conflictive notions of culture and cultural identity and difference, as they are currently employed in global and cultural studies. It should, however, avoid both the pitfalls of essentialism—which Huntington's study indirectly reveals to be another instrument of a will to power that regards itself as natural and universal—and the pitfalls of a politics of cultural identity/difference, based on an agonistic mentality. To this end, one can once again turn to the notion of globality as an infinitely diverse expression of the global aspiration. In this light, various cultures can be seen as primary modes in which the human desire for world making and self-fashioning, that is, the creative imagination, manifests itself.

Specific mentalities or modes of thought and behavior generate, and are at the same time generated by, specific ways of life, language, sound and image patterns, knowledge, art, architecture, institutions, as well as interactions with other human beings and with the physical environment—a complex and fluid web of causally reciprocal interdependence of physical, manufactured, and imaginary objects and events that can be called culture. Indeed, keeping in mind the essentialist caveat, it would be more appropriate to speak of cultures, rather than culture, whether local or global.

Every culture has the inner potential to renew or transform itself primarily through the imagination and its creative forms, including myths, narratives, folklore, artistic productions, ritual, and so on. In this respect, it is counterproductive, if not misleading, to draw a distinction, as Appadurai does, for example, between traditional and modern imagination, with one doing less social work and being less collective or less emancipatory than the other (Appadurai 1996, p. 5). There are many local collective imaginations that devise ever fresh ways of doing social work, based on the traditional and nontraditional creative resources of a specific culture. These local, creative resources may, moreover, fruitfully interact with similar resources from other, nearby or remote cultures. And there are many human factors other than power that motivate the various collective imaginations, such as playfulness, curiosity, generosity, love and care for others, aspiration toward personal development, spiritual transcendence and self-transformation, to mention only a few.

As I have already suggested, genuine transformations in mentality, whether local or global, happen very slowly, and cultural approaches to globalization that advocate power contests, conflict, and resistance can hardly help bring about such transformations. One should remember that theories have a performative, not just an explanatory or descriptive, value and therefore involve the ethical intervention of the theorist in the historical processes he or she is supposed to study or reflect on. Ultimately, then, the question is the old Socratic one: How productive is any particular theoretical model in creating sociocultural blueprints for sustainable human development? Models of order and disorder, conflictive identity and difference, aggression and resistance, master and slave races, autocratic rule and liberal democracy, have been around, in one form or another, for a long time and are integral to a certain mentality or mindset that has arguably led to the current human predicaments. So, assuming that we as human beings are genuinely committed to sociocultural paradigmatic shifts around the globe, are there alternative theoretical models that might be useful to insert into the local–global, intercultural dialogue at this historical juncture?

I shall have occasion to explore this question at some length in part II below, when I shall consider the general systems' view of the reciprocal causality between nature and culture, as well as the Buddhist, Taoist, and Sufi ways of reorganizing worlds (and cultures) within reference frames other than power. In the remaining pages of part I, I would like to propose a few basic principles that should guide us in building alternative cultural theories and practices of globalization. These theories and practices may, in turn, facilitate the emergence of a transdisciplinary field of intercultural studies and action, oriented toward global intelligence.

1 Local–Global Cultures and Subcultures

Within a global reference frame, we would be well advised to consider various cultures in an intercultural, comparative perspective, that is, to examine our local (Western) notions of culture in relation to their counterparts in other regions of the world. We should also attempt to arrive at a worldwide consensus about what it would mean to develop a global culture, or at least local–global cultures and subcultures, and what their founding principles would be. These and other, related issues would constitute the proper object and objective of a reconstructed field of intercultural studies.

Intercultural studies would hopefully transcend the current ideological and political impasse of Western-style cultural studies, at the same time that it would preserve and reorient some of its valuable insights toward global intelligence. It would situate itself in the vanguard of a much needed, comprehensive study of and sustained dialogue among world cultures, not only from a local, national, or regional perspective, but also from a global one. This type of intercultural learning and research project would necessarily involve an unprecedented, collective and cooperative effort on the part of learners, educators, scholars, researchers, and other practitioners from academic and nonacademic fields throughout the world.

Robertson's concept of global culture would be a good starting point for a discussion of the principles that could inform such an ambitious, transdisciplinary and intercultural project, oriented toward global intelligence. Robertson notes that our relatively recent commitment to the notion of a culturally cohesive nation–state "has blinded us to the various ways in which the world as a whole has been increasingly organized around sets of shifting definitions of the global circumstance. Yet the notion of a global culture should be just as meaningful as the idea of national–societal, or local, culture" (Robertson 1992, p. 114).

The concept of global culture can be very useful precisely because it has both contemporary and age-old dimensions. Robertson estimates, for example, that the current "globe talk" or the discourse of globality itself is a vital component of contemporary global culture. In his view, this discourse largely consists of the shifting and contested terms in which our world is defined as a whole. At the center of global culture one can find "images of world order (and disorder)—including interpretations of and assertions concerning the past, present and future of particular societies, civilizations, ethnic groups and regions" (ibid., p. 113). One should add, however, that if this "globe talk" were to become genuinely effective, it would soon need to move from its current oppositional or conflictive phase to a dialogical, nonviolent one.

One should, in turn, take into consideration the countless efforts of creating global cultures in the past. Robertson points out that, throughout human history, civilizations, multicultural empires, and other large social or political entities such as the modern nation–state have often been confronted with the problem of responding to an ever wider and, at the same time, increasingly compressed global framework. A theory of global culture should therefore also examine the ways in which such entities have sought to learn from others, without losing their sense of identity or, alternatively, have attempted to isolate themselves from cultural contact.

Another crucial aspect of global culture is the idea of humankind itself, which goes at least as far back as Karl Jaspers's Axial Age, during which "the major world religions and metaphysical doctrines arose, many centuries [indeed, millennia] before the rise of national communities or societies" (Robertson 1992, p. 113). Robertson appropriately places this idea among his tetrad of "major global components" (together with societies, individuals, and international relations), implying that at some future historical juncture it might again achieve the kind of world prominence that it had during the Axial Age. Of course, according to the proponents of the Aquarian Age and other members of the global "consciousness circuit" (Roszak 1975), we have already reached that phase—a social optimism that in the present global circumstance remains largely wishful thinking. This is not to say, however, that such thinking could not, under certain conditions, be socially effective or become a global reality.

One can further develop Robertson's concept of global culture in line with the nonlinear notions of globality and locality that I have proposed in the previous chapter. To begin with, one may amend "global culture" to local–global cultures, since there have been many globally oriented cultures and subcultures on our planet over the centuries, grounded in a wide variety of values, beliefs, and social goals and practices. One may thus continue to create new local–global cultures and subcultures, in which globally oriented local communities will interact with each other in cooperative, peaceful, and mutually enriching ways. Such local–global communities could revive or reinvent their own core of values, beliefs, and shared experiences that would be defined as human, in addition to their ethnic, national, and other features.

But, the very definition of "human" (and "nonhuman") should, in turn, become an object of sustained intercultural research and dialogue. The ideas of what constitutes humanity, what are its limits, possibilities, and destiny vary as widely in different cultures and religions of the world as those of time, space, and globality, and therefore cannot be taken for granted in a global reference frame. They would thus constitute a most appropriate starting point for a fresh kind of globe talk. One could also speak of "global citizens" and "global citizenship," although these phrases should perhaps also be amended to "local–global citizens and citizenship" and should equally become part of a sustained intercultural research and dialogue.

The local–global communities of the future would obviously not be devoid of "clashing of perspectives" (Featherstone 1995), any more than any previous local communities were. But these differences need not become linked to issues of mimetic identity, leading to Huntington's clashes of civilizations. One may delight in and encourage cultural differences, in order to enhance the richness and diversity of life, rather than use them as pretexts for violent conflict. One might also imagine new ways in which humans can negotiate their various identities. One could, for example, develop the notion of nonconflictive multiple identities, in which humans can easily and naturally identify not only with their community, ethnic group, nation, race, culture, civilization, and so forth, but also with other humans, as well as with all other beings on this planet and beyond.

One may finally note that the "ethnic core" and "common language" that Mike Featherstone, among many others, sees as prerequisites for some, if not most, local communities are not conditions sine qua non for all such communities (Featherstone 1995). In any case, ethnic factors and common language are far from constituting impassable barriers—on the contrary, they can be very helpful—in constructing a common human core of values and beliefs. In this connection, cross-cultural border villages, cities, and regions from around the world, which have not yet been sufficiently studied, and hardly at all from a transdisciplinary and intercultural perspective, can play an important role in imagining the local–global communities of the future.

Local–global communities would explore the common human core that binds them, rather than focus on the elements that separate them. In this respect, one may turn to those "commonalities of [human] civilization" that even a *Realpolitiker* like Huntington acknowledges as highly significant, at the very end of his study: "If humans are ever to develop a universal civilization, it will emerge gradually through the exploration and expansion of these commonalities. Thus, in addition to the abstention rule and the joint mediation rule, the third rule for peace in a multicivilizational world is the commonalities rule: peoples in all civilizations should search for and attempt to expand the values, institutions, and practices they have in common with peoples of other civilizations" (Huntington 1996, p. 320).

One should beware, however, of Huntington's political rhetoric of "rules," even though joint mediation (between various cultures and cultural concepts) may constitute a useful global principle and practice. In turn, the "rule" or, rather, the principle of abstention from violence could be given a positive formulation, free of thou-shalt or thou-shalt-not definitions of morality that rarely seem to be effective in practice. Such a principle of nonviolence (for which the English language does not even have a positive name yet) would be a fundamental one for any mentality of peace.

One should also beware of Huntington's notion of universal civilization, which he distinguishes from civilizations and opposes to "barbarism" (ibid., p. 320–321). Binary oppositions such as civilization and barbarism are emotionally charged terms, with a long and troubled world history, and will only recreate and perpetuate the mentality of power to which Huntington and other cultural theorists subscribe in various degrees. Rather, one should look for those local–global identifications and identities that are not based on the mimetic mechanism of "us against them," whether "they" are another ethnic group, culture, civilization, or an alien form of intelligence.

2 Intercultural Resonance and Responsive Understanding

In addition to the notion of cultural mimesis or imitation, and as a partial, constructive alternative to it, one can develop the concept of resonance, understood in both its physical and musical senses. I shall have more to say about this concept in part II below, when I explore its links to the general systems notions of reciprocal causality and amplifying feedback loops. For the moment, let me suggest that if various world communities have adopted and transformed ideas and practices from other cultures throughout human history, it is not because these cultures appear superior or more advanced, or as Huntington would say, more "powerful" than their own, thereby constituting authoritative models, worthy of imitation or emulation. Rather, it is because some of these ideas and practices often resonate with the inner being and collective imagination of receiving or responsive world communities, occasionally touching on their deepest desires of (self-)transcendence.

For example, one can argue that most cultural ideas and practices, including what the German Romantic thinkers called *Zeitgeist* (spirit of the age), propagate through resonance, rather than imitation. In fact, as soon as they begin to be perceived as imitations, especially in the modern age, they become problematic and lead to conflict, resistance, and rejection. One may speculate that what appear as cultural mimetic phenomena in romanticism and modernity often are improper or violent forms of resonance that tend to arrest, rather than to stimulate cultural production. In this regard, one may draw a distinction between a mimetic or destructive form of resonance, and a constructive or productive form. Both operate through amplifying feedback cycles, but the mimetic type amplifies conflict and violence, whereas the constructive one amplifies mutually nourishing, symbiotic social behavior.[1]

On the other hand, one should point out that the concept of imitation becomes counterproductive precisely with the advent of romanticism, which denies its constructive role, as resonance, in cultural interaction. The traditional doctrines of imitation, according to which the moderns imitated the ancients, emphasized the positive, learning aspects of the imitative process, on the authority of Aristotle's *Poetics*. At the same time, they attempted to contain its potentially violent, amplifying aspects, or mimetic resonance, within the linear hierarchical structure of the relation between model and copy, much as the central authority of a powerful king sought to arrest violent conflicts among warring feudal lords.

Once the Romantics presumably emancipated the relation of model and copy from its hierarchical, authoritarian framework, they also released the violent amplifying potential of imitation as resonance. According to the principle of reciprocal causality, mimetic resonance will eventually reach dangerously high levels, just as, in the absence of a powerful kingly authority, smoldering feuds among warlords tend to escalate and erupt into generalized violence. Revolutions can equally be seen as such amplifying feedback cycles whose potential violence is released and propagated through mimetic, rather than productive resonance. It is not a matter of pure coincidence, therefore, that violent revolutions became a common way of resolving social conflict during the Romantic period, once the divine authority of kingly power was abolished or contested.

Resonance can further be seen as operating at many different cultural levels, from the mere surface to the deep structure of a specific culture. For instance, the rapid spread of fast food chains around the globe need not be seen as a mimetic phenomenon produced by North American consumerist models, but as a superficial cultural practice that resonates, at the surface level, with a fairly common type of human behavior. No North American company could persuade anyone in the so-called Third, Second, or even First Worlds to buy and consume its fast food products or to create similar fast food establishments, if they did not resonate with certain (superficial) needs in the consumers, including a need for novelty.

Furthermore, it obviously takes much less time and effort to prepare and eat an industrial, ready-made meal rather than a homemade one. So, many of us are willing to sacrifice, for the sake of expediency or convenience, not only nutritional quality, but also that feeling of sacred ritual which comes from sharing a painstakingly and lovingly prepared meal and which is now mostly reserved for special holy days. In a sense, then, "America" and "American imperialism" are often used as a convenient and unreflective shorthand (and scapegoat) for our own improper drives or habits. Consumerist America, along with the other America of unlimited human possibilities, is already inside—and resonates with—all of us:

by identifying it as a source of ills or blessings we merely identify that part in us which we wish to change or, alternatively, to develop.

The concept of responsive understanding, closely related to that of resonance, could also be helpful in developing productive, local–global, intercultural exchanges. The term "responsive understanding" belongs to the Russian thinker Mikhail Bakhtin (1981), who initially uses it to describe Dostoevsky's approach to his characters and fictional universe. Bakhtin also calls this responsive understanding "active," "creative," and "dialogical." Significantly, he attributes it to literary discourse and distinguishes it from the monologic understanding of other types of discourse, such as the scientific and the political. Responsive understanding means watchful listening and empathetic, interactive participation in an ongoing dialogue, in which each participant carefully and lovingly preserves the integrity of the other.[2] In a sense, then, responsive understanding establishes the conditions of the possibility of productive cultural resonance and rechannels its mimetic, violence-amplifying potential.

One should add that responsive understanding also conveys the idea of responsibility, understood not as the thou-shalts and shalt-nots of conventional morality, but rather as a free and generous response to the calling of the other, interpreted as a principle of (self-)transcendence, such as God, globality, the human community, another human being, or the physical environment. It is not enough to put oneself in other people's shoes and understand or sympathize with their views, without getting involved in an effective manner. Therefore, responsive understanding can never be separated from willingness and ability to take positive, responsive action, the causally reciprocal effects of which will then propagate by amplifying feedback loops or resonance through the entire social system.

Even from this very brief discussion of responsive understanding, it ought to be clear that Bakhtin's term has relevance far beyond the realm of literature, at the same time that it highlights the value that literature and literary studies—indeed a literary or an artistic mode of thinking in general—may have in the present global circumstance. Many of the problems that face the contemporary world are questions of (intercultural) interpretation, translation, and communication, and our arts and humanities have already been dealing with these questions for a long time. So, artistic ways of understanding and dealing with the world may prove to be useful in political, economic, and sociological contexts as well.[3]

3 Global Awareness and Intercultural Comparative Analysis versus Cultural Critique

We may also wish, from the perspective of an emergent ethics of global intelligence, to develop alternative concepts to those of "cultural critique" and "critical thinking"—two of the most cherished notions of modern Western thought and the methodological cornerstones of current, Western-style, cultural studies. Precisely because of their sacred cow status, the Western scholarly community should submit them to both an extensive cultural anthropological study and a thorough self-analysis from a local–global perspective, rather than continue using them automatically, as if they were obvious, universally shared values. As I have suggested in the case of the post-Marxist critical stances, we particularly need to explore the ways in which the idea of critique has, throughout its history, contributed to a perpetuation of various disciplinary mentalities not only in the Western world but in other worlds as well. In this respect, even notions of cultural resistance, opposition, subversion, and survival are part of the arsenal of power that perpetuates politics as usual, be it in a "colonized" or "decolonized," local or global, environment.

From the standpoint of global intelligence, it would be wise to refrain from practicing cultural critique and critical thinking within an intercultural framework. These are local notions that might do well in certain Western intellectual circles, but not so well in other, nonwestern, circles. In the latter circles, they are often perceived as needlessly and counterproductively confrontational and aggressive, if not as "soft power" instruments of furthering Western, "imperialist" designs on other cultures. Although we probably cannot for the time being abandon them altogether, because they are too deeply ingrained in our mentality, we should become aware of and use them only in their proper, local, cultural contexts (as I have at least tried to do in the present study).

Even more importantly, we should consider other candidates to replace cultural critique and critical thinking as conceptual tools in a global learning environment. Such candidates in English might be "global awareness" (and "self-awareness"), "global attentiveness," and possibly "global consciousness," although the term "consciousness" has also accumulated a rather heavy philosophical and ideological baggage in Western intellectual history. These and other terms should, however, preserve the connotations of reflection and self-reflection that are necessary for any creative thinking and action that seeks human (self-) development. On the other hand, such terms should also remain free of the oppositional or agonistic connotations of "critique," as well as of its etymological link to "crisis," another pet concept of Western modernity. They would thus be less easily co-opted as instruments of power by any warring Western or other ideological factions in a global environment.

Intercultural comparative analysis should in turn replace cultural critique as the cornerstone methodology in a global milieu. But, such comparative analysis would need to be much refined and attuned to this milieu, as we have seen particularly from Huntington's work. As cultural anthropologists Steven Marcus and Michael Fischer, for instance, propose, we should develop "new practices of comparative analysis not among self-contained cultures but across hybrids, borders, diasporas, and incommensurable sites spanning institutions, domiciles, towns, cities, and now even cyberspace" (Marcus and Fischer 2000, p. xxix).

In intercultural comparative analysis, one may also employ such anthropological, literary, and rhetorical techniques as "dynamic, nonreductive juxtapositions," or "orchestrated engagements" of cultural horizons (ibid.). For instance, Michael Fischer and Mehdi Abedi imaginatively experiment with such techniques in a volume on *Debating Muslims: Cultural Dialogues in Postmodernity and Tradition* (Abedi and Fischer 1990). We should also consider the impact that reductive approaches in intercultural comparative analysis and their symmetrical opposites, pluralistic or relativistic approaches, may have within a global reference frame. Both types can be equally counterproductive in an intercultural environment: the first, because it may amplify cultural conflict and violence (as we saw in the case of Huntington); the second, because it may stall and eventually lead to a breakdown in intercultural dialogue.

One may note that, in addition to the new sites for intercultural comparative analysis mentioned by Marcus and Fischer, one should devote renewed attention to "self-contained" or historically constituted cultures as well. In this respect, intercultural comparative analyses should start from a secure, nonmimetic sense of cultural identity, rather than an insecure, mimetic one. It is from the same secure sense of identity, moreover, that members of diverse cultures should engage in intercultural dialogue and negotiation, which otherwise would be stalled by mimetic conflict and violence. An intercultural approach oriented toward global intelligence would obviously not require denying or repressing one's cultural position and identity, as extreme, North American, liberal forms of political correctness ask Westerners to do, playing on their sense of collective guilt—as if this were desirable or even possible.

Nor would such an intercultural approach require that humans ultimately develop a universal language, common to all humankind—this would be reenacting the myth of the linguistic paradise before the Tower of Babel and would most likely turn our planet into an utterly boring place. (Incidentally, the intercultural project of the Tower of Babel can be regarded as the paradigm for an endless series of unsuccessful human attempts at globalization.) Rather, a globally intelligent approach would require that researchers engaged in an intercultural project, and then gradually other members of the cultures involved in intercultural learning experiments, view their cultural position and identity through the others' perspectives, as well as their own.

One might thus begin by thoroughly exploring, in a collective, dialogical, and nonconflictive manner, the actual and imagined differences that may exist among the diverse human languages, cultures, value systems, and beliefs, which in turn determine the ways of thinking, feeling, and acting of various people. It might very well turn out that many of the clichés that various members of one culture circulate, automatically or intentionally, about other cultures and their members are based on incomplete knowledge or received opinion, as we have seen in the case of Huntington's and Payne's perpetuation of old Western (largely Christian) prejudices about Islamic cultures.

4 Intercultural Contact and Liminality

The notion of cultural contact is another conceptual tool that could prove helpful in intercultural comparative analysis, dialogue and negotiation, once we emancipate it from its current disciplinary and interdisciplinary contexts. For instance, we should distance it from the agonistic, confrontational approach that has been extensively employed in the North American culture wars, as well as in the so-called freedom movements of Latin America, Africa, and other parts of the world. One good example of this counterproductive approach is Mary Louise Pratt's notion of "criticism in the contact zone."

Pratt defines such zones as "social spaces where disparate cultures meet, clash, and grapple with each other, often in highly asymmetrical relations of domination and subordination" (Pratt 1992, p. 4). This notion might have been appropriate for past colonial and postcolonial contexts with their disciplinary discourses of struggle, opposition, domination, and liberation that have, in this last century, been repeated ad nauseam throughout the world. These discourses are all part of the arsenal of a power-oriented mentality and are counterproductive in the present global circumstance, in which antiglobalization movements are proving to be as ineffective, and as co-optable, as their earlier avatars.

Whereas one should certainly explore and find appropriate solutions to alleviate tensions in the troubled zones of intercultural contact, one should also study those zones in which people from various cultures are currently living—or used to live—side-by-side in peace and harmony. These studies, if conducted in a thoroughly cross-disciplinary and cross-cultural fashion and within their proper historical contexts may help reveal unexpected, nonlinear causes of conflict, as well as models of peaceful coexistence that might also be useful in other parts of the world. Indeed, it might well turn out to be the case that the no man's land between cultures, the empty spaces between borders, or the gray areas in which nothing is quite settled and in which new patterns of organization can gradually or suddenly emerge, may constitute privileged sites for intercultural dialogue and negotiation, rather than privileged sites of conflict. In this regard, we may redefine the notion of cultural or, rather, intercultural contact by linking it with the notion of liminality, which can also become an important conceptual tool for intercultural studies, oriented toward global intelligence.

I shall explore the concept of liminality at some length in parts II and III below. In the present context, I would like to initiate this exploration by distinguishing the liminal from the marginal as a locus of intercultural contact. Marginality can best be viewed in terms of a fluid and reversible dialectic of center and periphery, and as such it belongs to a vocabulary and pragmatics of power. A margin cannot function without a center, any more than a center can function without a margin. The center may wish to push its margin further and further away, because this margin always threatens to destabilize it and may eventually replace it. But, in the end, the center can define itself only in relation to the margin.

As Kant was well aware, power can emerge or manifest itself only if it meets obstacles or "opposition," such as the proverbial "barbarians at the gates," invoked by Huntington, Hardt, and Negri (2000), and many other ancient and modern cultural theorists. Stated differently, center and margin are in a relationship of mutual causality. A margin can be temporarily liminal, as in the case of a carnival, or a scapegoat ritual, or a social revolution. During such liminal events, when the margin temporarily replaces its center, the deliberate, temporary transgression of established social and cultural norms, if not closely monitored and contained, threatens to annihilate the whole power system. However, a properly managed ritual (or for that matter, social revolution) will always attain its objective of reinforcing the center, even as it may redefine it in the process. Academic disciplines and other power-oriented fields of knowledge equally follow this power pattern, as I shall show in part III of this study.

Liminality, on the other hand, may be present both inside and outside a power system. In this regard, a margin can be liminal, but a limen cannot be marginal. The limen (which in Latin means "threshold") can be situated between two or more power systems, but belong to none, as is the case, for instance, in the no man's land between two or more state borders. Again, a border, just like a margin, can be (temporarily) liminal, but a limen cannot constitute a border. Thus, liminality can both subsume and transcend a dialectics of margin and center. Unlike a margin or a border, it may not lead back to a center. On the contrary, it may lead away from it in a steady and irreversible fashion.

Because of its ontological ambivalence or apparent emptiness, liminality is viewed with suspicion by any power system, which always threatens to move into a space that it perceives to be void. But as soon as power moves into a liminal space, this space becomes a border or margin, and the limen moves outside it. That is why power is both horrified and fascinated by the liminal, which it conceptualizes as "nothingness" (being always outside its reach). Liminal spaces between cultures could, therefore, be the most appropriate sites for intercultural dialogue and negotiation: power views them as ultimately "unreachable," but can also view them, less threateningly, as "neutral" spaces where it can further its own objectives as well.

Liminal interstices, then, may be found both within compatible or incompatible worlds, such as those belonging to a mentality of power, and between incommensurable worlds, that is, between the various global subcultures generated by a power-oriented mentality and those belonging to an irenic one. The liminal interstices between incommensurable worlds or global subcultures are of particular interest in the present study, and I shall return to them in subsequent chapters.

5 Transdisciplinary versus Interdisciplinary Knowledge and Learning

First, one should point out that disciplinary principles and practices will obviously not work for a reconstructed field of intercultural studies, because they will only continue to aggravate the intercultural misunderstanding, conflict, and distrust that largely characterize the global arena today. Interdisciplinary principles and methodology will not do either, because these principles are entirely codependent with disciplinary ones. As a field of study and practice, therefore, intercultural studies would be cross-cultural and transdisciplinary and would employ a local–global approach to knowledge, oriented toward global intelligence. Its goal would be to remap traditional knowledge, as it is acquired and transmitted by various academic disciplines, be they scientific or humanistic, as well as to generate new kinds of knowledge within a cross-cultural and global framework.

A field of theory and practice such as transdisciplinary intercultural studies would need institutional arrangements that are different from the ones currently in place in our world of learning and research. One arrangement under which this field could prosper would consist of networks of cross-disciplinary and cross-cultural research teams, whose members would be selected from participating, academic and non-academic research institutions, located in various parts of the world. These teams would continuously change their disciplinary composition and research focus. They would constitute themselves not according to academic fields of study, but according to concrete, complex problems of a social, political, cultural, economic, medical, environmental, legal, military, or other nature that need to be solved at the local–global level in various parts of the world.

The overall mission of the transdisciplinary teams would be to study, produce, and apply local–global knowledge in a globally intelligent way. The teams would require a worldwide, cross-disciplinary and cross-cultural cooperative interaction among scholars, researchers, and practitioners in any field of human endeavor. In choosing the right members for such teams, it would matter less whether an individual researcher is a specialist or a generalist, an academic or a nonacademic, than whether she is open to ways of knowing and doing things beyond those she has been trained in and accustomed to.

The research programs for this field could be organized and conducted within six broad areas that are and will remain crucial for intercultural research and knowledge production in the foreseeable future: (1) globalization and local strategies for human development; (2) food, nutrition, and health care in a global framework; (3) energy world watch for sustainable development; (4) world population movement and growth; (5) information technology, new media, and intercultural communication; and (6) world traditions of wisdom and their contemporary relevance.

This list makes it clear that intercultural studies would call on all of the three major branches of learning (social sciences, humanities, and physical sciences) to participate in developing concrete transdisciplinary and cross-cultural projects in these six general areas. Each project would require different intercultural research teams and would select, on an ongoing basis, researchers and practitioners from those fields that would be of most assistance in attaining its particular research objectives. The essential feature of these teams would, however, be that specialists and nonspecialists would work together on resolving concrete human problems. Consequently, the principles of cross-cultural and cross-disciplinary dialogue and cooperation with a view to benefiting the entire planet, not only certain regions or interest groups, would necessarily constitute the ground rules for all research projects in intercultural studies.

In chapter 4 of *Remapping Knowledge*, I outline a detailed blueprint of a combined bachelor's and master's degree program in intercultural studies, based on the foregoing principles. In turn, in the appendix of the present book, I propose a pilot research program in intercultural knowledge management, designed as an advanced version, or continuation, of the first program. I also outline a concrete blueprint of a local–global institutional framework appropriate for both programs. Thereby I wish to demonstrate that the fundamental theoretical principles that I discuss in this book are not merely "pie in the sky." In fact, they can be implemented relatively easily (and cost-effectively) even under the present state of global education.

6 Democracy, Nonviolence, and Irenic Mentality

Finally, the concepts of globality, locality, the local–global, nonconflictive multiple identities, intercultural and global self-awareness, resonance, intercultural responsive understanding and dialogue, intercultural comparative analysis, intercultural contact, liminality, and intercultural studies oriented toward global intelligence would best achieve their full meaning and effect within a mentality of peace. Much has been written lately on the ideas of peace and nonviolence, especially in a political context. These ideas are usually seen as going hand in hand with, and are often subsumed to, the concepts of liberal democracy, human rights, social and gender equality, and the rule of law. A good example of this approach can be seen in a recent collective volume published by UNESCO under the title *From a Culture of Violence to a Culture of Peace* (1996).

In the introduction to this volume, Janusz Symonides, director of the UNESCO's Division of Human Rights, Democracy, and Peace, and Kishore Singh, program specialist in the same division, state that "democracies, as proved by historical experience, not only do not make war against each other but also through their systems of governance—rule of law, participation, transparency and accountability—diminish considerably recourse to violence" (Symonides and Singh 1996, p. 10). The editors further contend that, generally, there are two concepts of peace: the "negative, narrow understanding, reducing peace to a mere absence of war; and the positive, defining peace as a lack of war often enriched by further elements and guarantees which make peace constructive, just and democratic" (ibid., p. 15).

Symonides and Singh approvingly cite the Yamoussoucro Declaration (issued by the UNESCO International Congress on "Peace in the Minds of Men," Ivory Coast, 1989), which proposes to base a peace culture on "the universal values of respect for life, liberty, justice, solidarity, tolerance, human rights and equality between women and men" (ibid.). They believe this list to be reasonably complete, if "we add democracy, development, burden-sharing and responsibility as well as non-violence and peaceful resolution and transformation of conflicts" (ibid.). Symonides and Singh also support Hans Kung's proposal, in his essay on "A Planetary Code of Ethics: Ethical Foundations of a Culture of Peace" included in the same volume, for the elaboration and institutionalization of "world global ethics" (ibid., p. 129). According to the editors, this ethics would be based on "four irrevocable directives, requiring commitments to a culture of non-violence and respect for life; to a culture of solidarity and a just economic order; to a culture of tolerance and a life of truthfulness; and to a culture of equal rights and partnership between men and women" (ibid., pp. 19–20).

Symonides, Singh, and some of their colleagues share the view that liberal democracy, more than any other system of government, is conducive to peace and nonviolence, and that these concepts, moreover, thrive best in a democratic political system. This view is rather widespread in the West and has been promoted by Francis Fukuyama (1992), David Held (1995), and many other advocates of cosmopolitan democracy (Held's term). What historical experience seems to prove, however, is not that democracies are more peaceful and nonviolent than other systems of government, but that all such systems will promote and perpetuate themselves by whatever means they can muster, including (state-organized) violence. The fact that democracies seldom make war against each other proves only that they will form alliances with their own kind and will, if necessary, go to war only with other, incompatible systems of government. The latter will, in turn, seldom go to war with their allies, but only with perceived political enemies. During the 1960s and 1970s, for example, there were no greater fighters for peace than the socialist democracies of Eastern Europe and the Soviet Union, because it was in their interest (as it was in that of the Western-style democracies) to preserve the postwar political status quo.

All of the existing political systems, including the liberal democracies, are based on various mentalities of power and will resort to violence whenever they perceive their core values, beliefs, and interests to be seriously threatened. This is obvious, for example, in Symonides and Singh's implicit, unexamined distinction between "legitimate" and "illegitimate force" (1996, p. 17), or in Emmanuel Decaux's oxymoronic phrase, "struggles against a culture of hate" (ibid., p. 55), in his essay on "Normative Instruments for a Culture of Peace" included in the same volume. Such distinctions sound rather hollow in view of the fact that, throughout past and recent history, most governments, whether democratic or not, have routinely invoked principles of legitimacy and justice in order to perpetrate the worst kind of violent acts against each other.

Equally hollow and tautological is the distinction between peace as a mere absence of war and peace as lack of war, guaranteed by democratic principles. Both types of peace depend for their definition on war, which Symonides and Singh call a "factual state, an attempt to solve disputes and conflicts and to achieve domination through armed force and violence" (ibid., p. 14). The subtle implication seems to be not that one should renounce domination altogether, but only that it should be achieved through means other than war, perhaps through the soft power mentioned by Nye and Huntington. If everything else fails, however, there is always legitimate force to be invoked and applied. The current international political situation, brought about by the September 11 events in the United States and the ensuing global war on terrorism, sadly illustrates the truth of this proposition. For our current world leaders, no less than for Symonides and Singh, the concept of peace remains tied up with and subordinated to power politics.

Conversely, it would take a radical shift in mentality in order to change the power politics that dominate the global environment at the present time. A mentality of peace and local–global cultures of nonviolence would therefore effectively be built on principles other than human rights, tolerance, nondiscrimination, social, racial, and gender equality, no matter how attractive and incontrovertible these democratic principles might appear at first sight. To be sure, within the present Western local contexts, these principles should continue to be regarded as valuable and praiseworthy, constituting proper political ideals for liberal democracy—and one may note that to date they have not progressed far beyond the ideal stage within these local contexts as well.

Within a global reference frame, however, Western democratic tenets often appear as improper expressions of the global impulse and are hardly a sound foundation for a "planetary code of ethics." Outside the Western world, they are frequently perceived, rightly or not, as yet another ploy of Western imperialism to impose its system of values and beliefs on the rest of the world and have often produced exactly the opposite effect from the one intended—that is, more conflict and violence. In this regard, Terry Eagleton's point is well taken that the alleged final triumph of Western-style capitalism throughout the planet "may also prove exceptionally dangerous for it" (Eagleton 2000, p. 82).

On the other hand, if some societies around the world have now begun to invent their own forms of democracy, it is not because these forms have been imposed or adopted from outside or because they are demonstrably better than others, but because certain democratic principles resonate with their current inner aspirations. This also means that there is no single formula for a successful democratic society, within or outside the Western world, that can be imported and replicated throughout the globe. Instead, each country will eventually find its proper social, economic, and political frameworks, should it choose to take a democratic path.

Nor should we content ourselves, if we wish to build genuine cultures of peace, with the lesser-of-two-evils arguments advanced in support of liberal democracy in relation to various forms of authoritarianism. Democratic principles, no less than authoritarian ones, are agonistic political principles and as such they will always perpetuate the very mentality they claim to oppose. As long as peace and nonviolence remain political instruments of power, for example, in pacifism or in the soft power of liberal democracy, they cannot achieve anything but the latter's goals. If one would truly like to build and promote a mentality of peace, therefore, one could do so only outside current politics and political systems.

What one would need to change in the first place is precisely the rather widespread practice of subordinating all values and beliefs to political, utilitarian goals. One would then imagine or reinvent principles and practices based on the idea of peace, defined not as lack of war, but as a state of mind and mode of behavior in which power ceases to be the organizing principle. But this would take a determined and truly unprecedented effort on the part of the collective imagination of humans all over the planet and can hardly be accomplished overnight or by a single world culture, no matter how well intentioned and globally oriented that culture might be. Meanwhile, we should by no means give up imagining such irenic principles and proposing them, in turn, for sustained, transdisciplinary, and intercultural research and dialogue.

II

An Intercultural Ecology of Science

3

Intercultural Dimensions of Natural Science

The principles and methods of a transdisciplinary field of intercultural studies, oriented toward global intelligence should become operative not only in the social sciences, but also in the physical and life sciences. The task of carrying such principles and methods over to the latter domains of knowledge is, however, made doubly difficult by the apparent worldwide success of what one may term Western-style, mainstream scientific theory and practice. Over the past century, this largely reductionist form of science has been catapulted into a globally visible cognitive paradigm, engaging in complex, amplifying feedback loops with the utilitarian mentality of industrial and network capitalism. The latter has, moreover, turned it into a faithful and obedient servant, under the guise of "technoscience."

The main intellectual driving forces behind this apparent global success of Western-style, mainstream science have, ironically, been the theory of relativity and quantum mechanics in physics and the theory of evolution in the life sciences. The disciplinary and interdisciplinary fields that have sprung up as a result of the institutionalization of these theories, especially elementary particle physics, genetics, molecular biology, and digital information technology, with their wide-ranging biotechnological applications, are significantly changing the nature of life on earth, for better or for worse. From the standpoint of an emergent ethics of global intelligence, however, we may wish to ask the following questions: Just how successful is the Western-style, mainstream scientific paradigm and, more to the point, how is this "success" defined? What are its positive and negative consequences for future human development? Are its negative consequences irreversible? Are there other viable ways of practicing science that would stimulate sustainable human development in a local–global learning and research environment?

Before attempting to answer any of these questions, it might be helpful to clarify what I mean by the "Western-style, mainstream scientific paradigm." I define this paradigm as an (inter-)disciplinary system of scientific theory and practice that was largely developed during the last century in the West, but that is based on a combination of four, much older, philosophical premises: materialism, objectivism, reductionism, and linear causality. According to these premises, reality is objective, or independent of (although knowable by) human subjectivity; it consists of material particles that can be reduced to the smallest entity; and it is strung together by cause and effect in linear and hierarchical fashion. This paradigm, moreover, postulates that scientific knowledge is objective, abstract, and universal; and that it is grounded in a handful of immutable physical laws that obtain uniformly throughout the universe and that should ideally be expressed in one simple mathematical formula.

In turn, according to the mainstream scientific paradigm, knowledge can be acquired, accumulated, and transmitted through a unique and privileged methodology, the experimental scientific method. This method is a highly structured and regulated, linear process of tracking cause and effect, an elaborate protocol of trial and error, experiment and critical (dis-)proof. It continually approximates, although it may never reach, universal and eternal truth, valid in all places and at all times. Finally, I call this kind of science a Western-style paradigm because it is equally operative in non-Western cultures, where it enters in resonance with similar local scientific traditions and produces similar constructive or destructive effects, reverberating throughout the planet. Indeed, it has created a global subculture of its own, which parallels and interacts with the global subculture of network capitalism, to which it has now become subordinated.

In this second part of my study, I shall first examine, in the present chapter, some of the main theoretical assumptions behind this scientific paradigm in terms of its ethical implications and practical consequences within a global reference frame. Even though Western-style, postmodernist critiques of mainstream, rationalist and reductive scientific discourse have done much to discredit the so-called metanarratives or grand ideological and pseudoscientific schemes (e.g., religious, Marxist, liberal, and neoliberal ones) of the nineteenth and the twentieth century, contemporary mainstream science has created its own neofoundational discourse that claims to marshal all knowledge under a single unified theory. I shall therefore examine two of these reductionist theories, namely "consilience" in sociobiology and, briefly, the "theory of everything" in elementary particle physics. I shall show the dangers of their dogmatic application in an intercultural, global environment, where they can all too easily be used to support and justify a renewed Western cultural expansionism, if not imperialism.

By contrast, I shall look at other available scientific models of evolution, outside the mainstream, which may be fruitfully adopted, refined, and developed in local–global contexts. I shall focus particularly on the local–global advantages of the nonlinear scientific models that are present in certain strands of general systems theory, complexity and chaos theory, and environmental science in the West. The main purpose of these analyses is to set up the preliminary conceptual framework for the development of an intercultural ecology of science, oriented toward global intelligence.

A crucial part of this project, moreover, would be the elaboration of an ecology of ecology that would propose ecological principles and practices appropriate for an intercultural, global reference frame. Some of these principles and practices are already present in the ancient traditions of wisdom throughout the planet, for example, in early Taoist thinking in China and early Buddhist thinking in India, as well as in the later Sufi tradition in Egypt, Persia, Turkey, and other Islamic countries. Therefore, in chapters 4 and 5 below, I shall also undertake an exploration of such thinking and its direct relevance to the project of elaborating an ecology of ecology.

1 Sociobiology in a Global Framework: An Intercultural Evaluation

In order to understand some of the global, intercultural implications of the mainstream, reductive scientific system of values and beliefs, one may turn to one of its most outspoken and articulate contemporary advocates, Edward O. Wilson. A Harvard entomologist turned sociobiologist, Wilson has written a number of books arguing for a biological interpretation of society and culture. His recent book, *Consilience* (1998), widely acclaimed in both scientific and lay circles, especially in the United States and England, can serve as an excellent case study for the universal claims of reductionist, mainstream science. Such claims are based primarily on Darwinian evolutionary theory (narrowly construed as the survival of the fittest through natural selection), and the analytic or reductive method in the physical and life sciences.

Wilson claims both the Presocratics and the philosophers of the Enlightenment, chiefly Thales of Miletus, Descartes, Condorcet, Bacon, Newton, and Goethe, to be his forerunners in the universal "dream of intellectual unity" (Wilson 1998, pp. 2–3, 13–36). For him, this dream takes the form of "consilience" among the three main branches of learning: natural sciences, social sciences, and the humanities. He borrows the term from William Whewell who, in *The Philosophy of the Inductive Sciences* (1840), defines consilience as a "jumping together" of knowledge by "the linking of facts and fact-based theory across disciplines to create a common groundwork of explanation" (Wilson 1998, p. 7).

According to Wilson, the principle of consilience is already implicitly at work in the natural sciences. He therefore proposes to use it for building a bridge from the natural sciences to the social sciences and the humanities as well. Throughout his book, he attempts to provide a blueprint for a general, consilient explanation of human nature, proceeding from the deep history of genetic evolution to modern society and culture. But Wilson, unlike Whewell, identifies consilience as a scientific method exclusively with reductionism and analytic mathematical modeling. He defines reductionism, which to him is "the cutting edge of science," not only as the "breaking apart of nature into its natural constituents," but also as the reduction of nature to simple, universal laws by mathematical artifice (ibid., p. 58). In turn, he sees "total consilience" as a strong form of reductionism that "holds that nature is organized by simple universal laws of physics to which all other laws and principles can be reduced" (ibid., p. 59).

Consilience is certainly a worthwhile goal to achieve, within and outside science, but the question is under what theoretical and practical frameworks. There is nothing wrong with the reductive or analytic method per se, as long as it is employed, under certain conditions, as an artifice to solve a specific, "local" scientific puzzle. The problems start cropping up, however, when it is equated with science in general and then raised, as such, to a universal principle of objective truth. Wilson does precisely that, for instance when he declares that there is only one type of consilience, in which all phenomena are "based on material processes that are ultimately reducible, however long and tortuous the sequences, to the laws of physics" (ibid., p. 297). He addresses an otherwise opportune call to researchers to treat the boundary between the scientific and literary cultures "not as a territorial line, but as a broad and mostly unexplored terrain awaiting cooperative entry from both sides" (ibid., p. 138). But then he turns consilience into a one-way street from reductionist biology to the other branches of learning, blurring the line between opportune and opportunistic.

Throughout his book, Wilson attempts to give the impression that all science is based on the same reductive principles.[1] But, contrary to what Wilson would have his readers believe, scientific holism is usually associated with organicism, and not with Cartesian or any other type of reductionism. In turn, organicism is a scientific tradition that is at least as old and venerable as reductionism, being traceable at least as far back as Pythagoras in the Western tradition, and even farther back in Indian and Chinese traditions, as we shall see in chapter 4 below. Its scientific proponents regard it as the opposite, not the complement of reductionism.

Another problem is Wilson's agonistic and hubristic intellectual attitude that undoubtedly stems from his evolutionary belief in the survival of the fittest, which he often interprets as the survival of the strongest. For example, he notes in relation to ants and their systems of communication: "In war—and Nature is a battlefield, make no mistake—one needs secret codes" (ibid., p. 76). It is this "natural" war of all against all that he extends to human culture and impels him to affirm the absolute supremacy of (reductive) science. He thereby engages, ironically, in a mimetic relationship with the various fundamentalist religious doctrines whose illiberality he decries elsewhere in his book. Wilson theorizes and practices what one may call a strong form of scientific fundamentalism (whereas most mainstream scientists practice, without theorizing, a milder form of it). His philosophical beliefs are similar to those of Huntington, who equally bases his thesis of the "clash of civilizations" on the antiquated evolutionary notion of the survival of the strongest. Wilson, in turn, extends Huntington's idea of cultural imperialism to the life sciences as well.

Wilson's fundamentalist claims for reductive mainstream science, no less than Huntington's claims for the necessity of (Western) cultural imperialism, are unhelpful in the present global circumstance, to say the least. From the perspective of an emergent ethics of global intelligence, they would be quaintly obsolescent or merely laughable, if they did not represent the tacit belief and practice of a majority of the Western scientific community, engaged in multibillion dollar research projects to fuel the formidable industrial, commercial, and military machines of the West. Wilson presses scientific reductionism into the service of the "culture wars" (Eagleton 2000), which first flared up in the Western intellectual community with the advent of postcolonialism and postmodernism, and which Huntington attempts to fuel as well.

Unlike Huntington, however, and yet as if to confirm his hegemonic cultural theses, Wilson considers the greatest divide within humanity to be not that between civilizations, ethnicities, or religions. Rather, it is that between scientific and prescientific cultures. According to him, Western mainstream science "is neither a philosophy nor a belief system" (Wilson 1998, p. 48). Rather, it is "a combination of mental operations that has become increasingly the habit of educated peoples, a culture of illuminations hit upon by a fortunate turn of history that yielded the most effective way of learning about the real world ever conceived" (ibid.). For Wilson the reductive scientific mindset and practices are the cornerstones of a future global civilization.

Wilson means no "disrespect" to "prescientific people" (ibid., p. 50), but they are simply unable to delve into the nature of reality, because they are unaided by the prosthetic instruments of contemporary science and are equipped with nothing beyond common sense. According to him, no human cognitive enterprise has worked "better" than Western mainstream science (ibid.). Indeed, the laws of physics are "so accurate as to transcend cultural differences. They boil down to mathematical formulae that cannot be given Chinese or Ethiopian or Mayan nuances. Nor do they cut any slack for masculinist or feminist variations. We may even reasonably suppose that any advanced extraterrestrial civilizations, if they possess nuclear power and can launch spacecraft, have discovered the same laws, such that their physics could be translated isomorphically, point to point, set to point, and point to set, into human notation" (ibid., p. 52).

According to Wilson, reductionism, "given its unbroken string of successes," may appear today as "the obvious best way to have constructed knowledge of the physical world" (ibid., p. 31). Yet it was difficult to grasp at the beginnings of science. For example, Chinese scholars "never achieved it" (ibid.). Although they had "the same intellectual ability as Western scientists," the Chinese focused on holistic properties and on the harmonious interrelation of entities (ibid.). In the Chinese holistic worldview, "the entities of Nature are inseparable and perpetually changing, not discrete and constant as perceived by the Enlightenment thinkers. As a result the Chinese never hit upon the entry point of abstraction and break-apart analytic research attained by European science in the seventeenth century" (ibid., p. 32). Western science, on the other hand, "took the lead largely because it cultivated reductionism and physical law to expand the understanding of space and time beyond that attainable by the unaided senses" (ibid.).

The preceding citations reveal Wilson's strategy of omission and exclusion that characterizes his reductive method in general. Reductionism may be useful as a scientific artifice, but can be highly suspect as a rhetorical device. Reductive intercultural comparisons are especially insidious and pernicious, as we have seen in the case of Huntington, because they are rooted in the limited, clichéd knowledge about other cultures presumably displayed by Western, particularly North American, lay audiences. Wilson scrambles and shuffles around his historical periods and dates—in this case ancient Chinese thought and seventeenth- and eighteenth-century European Enlightenment—in order to achieve the historical and cultural reduction needed to prove the superiority of Western science over its Eastern counterparts. And he does all of this under the guise of evolutionary theory and scientific objectivity.

To refute Wilson's spurious historical argument in favor of Western reductionism, one can point out, on the one hand, that the holistic and dynamic approaches of the ancient Chinese are certainly not absent from the ancient Western (Hellenic) tradition, for example in Thales and other Presocratic thinkers, whom Wilson opportunistically and mistakenly lists in the reductionist camp. These Hellenic thinkers, moreover, are neither materialists nor essentialists. They do not operate with concepts of identity and substance, or with object–subject dichotomies, as Wilson implies. Instead, not unlike the Taoists and the Confucians, they employ a holistic notion of reality as dynamic process, where phenomena (not entities) affect each other continuously in terms of reciprocal, rather than linear, causality. It is only with Parmenides, Plato, and Aristotle that a linear, essentialist, and substantialist worldview begins to emerge. In turn, Leucippus and Democritus can be considered, somewhat more plausibly than either Thales or the philosophers of the Enlightenment, as the first reductive thinkers in the Western tradition, through their theories of the atom.[2]

On the other hand, Wilson ironically proves the opposite of the point he wants to make, namely that the ancient Chinese were unfamiliar with reductive or analytic ways of thinking. He tries opportunistically to use Joseph Needham's work to document this alleged omission on the part of Chinese thinkers. Needham was a Western biologist who, unlike Wilson, was deeply committed to organicism and who became a historian of traditional Chinese science and civilization precisely in order to support his organismic concept of biology. Notwithstanding, Wilson cites him to the effect that the ancient Chinese "had a distaste for abstract codified law, stemming from their unhappy experience with the Legalists, rigid quantifiers of the law who ruled during the transition from feudalism to bureaucracy in the Ch'in dynasty (221–206 B.C.)" (Wilson 1998, p. 32). But what would one call abstract and rigid quantifiers of the law, if not reductionists, in any culture? Their case conclusively proves that the ancient Chinese, far from being unaware of reductive methods, decided at least temporarily to abandon them as counterproductive to their society.

Wilson continues unwittingly to undermine his own argument, when he observes that the "Chinese scholars abandoned the idea of a supreme being with personal and creative properties. No rational Author of Nature existed in their universe; consequently the objects they meticulously described did not follow universal principles, but instead operated within particular rules followed by those entities in the cosmic order. In the absence of a compelling need for the notion of general laws—thoughts in the mind of God, so to speak—little or no search was made for them" (ibid.).

One may plausibly argue, adopting for a moment Wilson's cultural evolutionary beliefs, that the fact that the ancient Chinese considered and then abandoned the idea of a supreme anthropomorphic deity shows that this idea was incompatible with a holistic approach to knowledge and was a step forward in the "evolution" of human thought. Wilson himself, at the end of the twentieth century A.D., abandons the idea of an anthropomorphic (Christian) God only half-heartedly so that, evolutionarily speaking, the Chinese scholars of the third century B.C. seem to be way ahead of him.[3] But Wilson wants to have his Western reductionist cake and eat it, too: although he argues, in this particular context, that the idea of a "rational Author of Nature," which the Chinese abandoned, made possible Enlightenment reductive thinking, in his account of the Enlightenment in the very next few pages of his book, he deplores the fact that most thinkers of that period were reluctant to give up the idea of a revealed divinity, albeit they invented deism to replace it (ibid., pp. 34–35).

These examples go to show that Wilson's strong fundamentalist claims for Western reductionist science constitute an improper form of globalism. His culturally tendentious aporiae dramatize the urgent need for extensive and credible intercultural research programs not only in the social sciences and the humanities, but also in the life sciences. Moreover, if these programs are to be effective, they need to be carried out by transdisciplinary and cross-cultural, collaborative teams, in order to avoid reductive, invidious, and misinformed intercultural comparisons that can only raise additional barriers to responsive (and responsible) intercultural communication and understanding. Of course, sociobiology should not be excluded from such research programs, but should be invited to come in as an equal partner, rather than as a supreme arbiter of scientific truth.

Wilsonian sociobiology should in turn abandon its hegemonic pretensions and undertake a thorough examination of its theoretical assumptions. The first evolutionarily "strong" idea that needs to go is precisely that of the survival of the fittest through natural selection, applied to human civilizations and interpreted as the survival of the strongest.[4] I would like to take one concrete example, among many, to show its arbitrary, subjective nature that invalidates Wilson's claims of objective, scientific truth. This example concerns Wilson's concept of epigenetic rules that guide cultural behavior and explain cultural change/evolution.

Epigenesis, according to Wilson, is "the development of an organism under the joint influence of heredity and environment" (ibid., p. 214). There are species-wide properties of human behavior that are guided by epigenetic rules. These are innate operations in the sensory system and brain that "predispose the individuals to view the world in a particular innate way and automatically to make certain choices as opposed to others" (ibid.). Above all, epigenetic rules "direct individuals toward those responses most likely to ensure survival and reproduction" (ibid.). Wilson repeatedly mentions fear of snakes as a universal epigenetic phenomenon. He contends that this fear is instinctual, that is, epigenetically prescribed, as a form of prepared learning. He traces it back to the fact that poisonous snakes account for many human deaths, so that ophidiophobia, or aversion of snakes, has survival value. Wilson then turns ophidiophobia into an epigenetic rule, claiming that it confirms his theory of gene culture coevolution (ibid., pp. 139–140).

Wilson provides a fascinating account of the presence of serpents as cultural symbols in many human societies (ibid., pp. 77–78, 84–88). But, his linear, causal biological explanation of this presence seems to me both trite and arbitrary. Zoophobia in human societies extends even to creatures that are as a rule unthreatening to humans, such as small, nonpredatory birds, mice, rats, certain nonpoisonous insects such as cockroaches, and so forth. There is little survival value in fearing these creatures. One could always argue, I suppose, that they are pests and disease carriers. But, then, most of nature can be seen this way, say, by neurotic urbanites or by the chemical, pest-control industry.

By the same token, ophidiophilia may be more extensive than Wilson believes and does not necessarily imply a "special way" of liking snakes (ibid., p. 86). Serpents, as Wilson himself points out, can be not only evil, but also benign cultural representations. They have often served as symbols of fertility, wisdom, and healing in cultures whose natural environments abound, or do not abound, in poisonous snakes. There are alternative explanations for their ubiquitous presence in the human imagination, which are not necessarily connected to biology or to ophidian potential deadliness to humans. They are more credible, however, because they are less reductive and more inclusive than their sociobiological counterpart.

Riane Eisler, for example, offers such an alternative explanation in *The Chalice and the Blade: Our History, Our Future* (1988), as part of a much larger cultural-historical thesis. Eisler argues, from a broad feminist perspective, for a "partnership society" in which men and women are equal partners in creating humanity's future. Her argument relies on recent anthropological and archeological research on prehistoric and archaic agrarian cultures, which were by and large matrilinear and matrilocal. She hypothesizes an archaic worldwide shift from a "partnership" to a "dominator" society, in which "androcentric" values such as violence, aggression, and competition prevailed over cooperative, nonviolent, and nurturing values, usually associated with women. She argues against calling these archaic partnership societies "matriarchies," because, according to her, they were not hierarchically organized, but egalitarian. Eisler believes, moreover, that the dominator societies that have prevailed so long in human history might now be coming to an end: they have become evolutionarily unstable, if not entirely self-destructive, and might again be replaced by partnership societies.

There are some problems with Eisler's argument, especially connected with the historical and archeological evidence she calls in to support her thesis of archaic partnership societies. Such societies were certainly less than ideal, as she herself partly acknowledges. For instance, despite Eisler's claims to the contrary, the evidence shows that many of these societies were in fact matriarchies. Some of them were more hierarchically organized than others, usually with a queen at the top of the social pyramid. They did employ force and violence to keep the social and religious fabric together, for instance in the form of human sacrifice, usually of young males who were both scapegoats and the queen's temporary "partners."

Nevertheless, these agrarian societies were more egalitarian and less prone to high-intensity violence than nomadic warrior societies. This was the case, perhaps, because farming entails hard, cooperative, communal labor, sedentary and stable habits that are patterned on the seasonal cycles, and a mostly meatless diet. As we know from extant traditional agrarian communities, meat is a scarce commodity and is as a rule consumed only on special occasions, such as religious holidays and other communal celebrations. In turn, hunting is a leisurely pursuit, mostly indulged in during the agricultural off-season. Domestic animals are primarily used for their by-products and for labor in the field, rather than for meat.

On the other hand, Eisler seems to me on safer anthropological ground when she suggests that our modern cultures are the result of a mixture between the archaic agrarian societies and the predatory, nomadic warrior societies that conquered and colonized them. The warrior societies partly imposed their might-makes-right mentality on the agrarian ones and partly adopted the latter's more quiet, cooperative values, especially in times of peace. It is also quite plausible, one might add, that these cooperative values were mostly expected of women, children, farm laborers, and other dependents, as befitting their social station and pursuits, whereas warlike, competitive values were mostly required of aristocratic male warriors. These social functions and divisions of labor could all too easily become blurred or lead to widespread inequality, abuse, and exploitation of the weaker members of the community, including the so-called weaker sex, as the troubled histories of many Western and non-Western societies clearly reveal.

But, independent of the historical record—or with careful, credible reconstructions of it—there is no good reason why we should not work toward the partnership societies envisaged by Eisler, even as some of us might balk at the term "partnership": this word evokes, at least to me, either North American law firms or the sexually neutral, politically correct, term "partner," of Californian extraction, denoting unmarried couples of all sexual persuasions. Be that as it may, Eisler's type of nonviolent, peace-loving, cooperative, and nurturing society seems to me a very attractive, sane, and sustainable alternative for human development, given the fact that, evolutionarily speaking, our "dominator" societies have not taken us very far in the past few thousand years, but quite the contrary.

To return to serpents and epigenetic rules, in many agrarian, matriarchal societies snakes were far from being feared as deadly enemies. Rather, they were regarded as benign symbols of fertility, wisdom, and healing, being associated with the earth and chthonic divinities. Indeed, in many archaic cultures, including those of Minoan Greece, they were benign divinities themselves. One may plausibly argue, as Eisler does, that only with the advent of the nomadic, warlike societies that brought their bellicose sky gods and goddesses with them, did the older chthonic divinities, such as serpents, become extinct, neutralized, or transformed into maleficent natural or supernatural forces (Eisler 1988, pp. 87–89). The fact that, to patriarchal warrior cultures, serpents were symbolic vestiges of an older, largely matriarchal social order, to be feared and distrusted (and therefore controlled and repressed), must have undoubtedly contributed to this transformation.

Did epigenesis reinforce, through ophidiophobia, archaic patriarchal fears of an even more archaic, matriarchal social order? It is hard, if not impossible, to tell. In any case, the theory I have just invoked to "explain" the complex, multivalent ophidian symbolism in human civilization is more elegant and economical than Wilson's epigenesis, because it does not appeal to mysterious "innate operations in the sensory system and brain" (Wilson 1998, p. 214), but deals with ascertainable cultural evidence, accounting, moreover, for its contradictory nature.

Interestingly, Wilson has his own version of the transition from partnership to dominator societies, although he does not use this terminology, nor does he speak of a civilizational shift, as Eisler does. Rather, he presents it as a linear "evolutionary" process. He raises the issue in relation to cooperation, conflict, and epigenetic fear of strangers. He suggests, not unlike Huntington, that the "complementary instincts of morality and tribalism are easily manipulated," especially as human civilizations evolve: "The rising agricultural societies, egalitarian at first, became hierarchical. As chiefdoms and then states thrived on agricultural surpluses, hereditary rulers and priestly castes took power. The old ethical codes were transformed into coercive regulations, always to the advantage of the ruling classes" (Wilson 1998, p. 283).

All of these developments are presumably part of cultural evolution directed by the survival of the fittest. Wilson invokes the animal world to argue for the survival and reproductive value of "dominance orders" (Eisler's equivalent of "dominator societies"). According to him, membership in a dominance order "pays off" for both the dominating and the subordinate animal, giving them better access to food, shelter, and better protection against enemies. Subordination in the group is not permanent, giving other animals a chance to dominate in turn, as they advance in rank and appropriate more resources. Wilson then concludes: "It would be surprising to find that modern humans had managed to erase the old mammalian genetic programs and devise other means of distributing power. All the evidence suggests they have not. True to their primate heritage, people are easily seduced by confident charismatic leaders, especially males" (ibid., p. 289).

Forgotten are the earlier egalitarian societies, which, Wilson appears to imply, must not have been as adaptive as the dominance orders, because they were less conducive to survival and reproductive success. He imparts an aura of inevitability and finality to these dominance societies, confirmed, according to him, by "overwhelming" evidence from the animal world. As he puts it in an earlier, related argument: "And so it has ever been, and so it will ever be" (ibid., p. 282). Yet, Wilson wants again to have his (survivalist) cake and eat it, too. Despite his implicit endorsement of dominance societies, he expresses hope that a different global social consensus will be reached, presumably under the dispensation of reductionist science, once it dispatches the evil serpent of religion: "No one can guess the form the [global social] agreements will take. The process, however, can be predicted with assurance. It will be democratic, weakening the clash of rival religions and ideologies. History is moving decidedly in that direction, and people are by nature too bright and too contentious to abide by anything else" (ibid., p. 285). Judging from Wilson's own account, however, we humans have so far been mightily contentious, but none too bright. Consequently we have, for thousands of years, abided by anything else but democracy. So it is by no means clear where Wilson's faith in democracy comes from, for surely it is not from his sociobiological theories.

Most mainstream scientists are no more ready than Wilson to give up the ideology of evolutionary progress and success that has supposedly served them so well. Of course, in their rare self-reflective moments these scientists see themselves, at least in print, as disinterested, selfless seekers and servers of objective knowledge and truth. Indeed, they see themselves as worshippers in the "Temple of Science" as Albert Einstein very aptly (and with no trace of irony or self-irony) puts it. In practice, however, those claiming to be in possession of the truth, or at least of parts of it, are stern, Cerberian gatekeepers to this new temple, and will exact a high price to let noninitiates and neophytes in. As I pointed out in the introduction, this is a manifestation of the famous equation of knowledge and power that has long been an explicit or implicit model of scientific thinking and practice in the West. Such disciplinary models of knowledge are deeply rooted in the European philosophical tradition. But, they also find their equivalents in the philosophical and/or religious traditions of other cultures, which partly explains why they have been exported so successfully, in the past century or so, practically all over the world.

I shall discuss at some length the disciplinary nature of our cognitive paradigms as a whole in part III below, where I explore the disciplinary and interdisciplinary models of our higher education. Here I shall limit myself to saying that the power/knowledge equation also explains the highly structured and ritualized nature of mainstream scientific communities (whether inside or outside academia), which can be analyzed as modern avatars of patriarchal tribal culture. Wilson largely denies this possibility, claiming that "oddly, there is very little science *culture*, at least in the strict tribal sense. Few rites are performed to speak of. There is at most only a scattering of icons" (Wilson 1998, p. 61; emphasis in the original). In the very next sentence, however, he confesses that one does "hear a great deal of bickering over territory and status. The social organization of science most resembles a loose confederation of petty fiefdoms" (ibid.). This confession seriously undermines Wilson's attempt to convey the impression of the unity of the natural sciences under the banner of reductionism, but confirms the analyses of certain historians of science who have shown that the history of the natural sciences is as contentious as that of the social sciences and the humanities.

Yet, Western mainstream physical scientists differ from other researchers, say, in the social sciences or the humanities, insofar as they have for the most part internalized clan discipline. They seldom wash their linen in public, largely because of the high economic and social stakes involved in their research. They present a united front to outsiders, according to a tightly organized and often repressive esprit de corps, similar to that of openly hierarchical, disciplinary organizations such as the military, police, and religious orders. Like members of these organizations, mainstream scientists can be castigated, or even excommunicated and neutralized by their peers, if they break ranks.

Even a self-avowed apologist of modern science such as Thomas Kuhn confirms this evaluation of the mainstream scientific subculture. In *The Structure of Scientific Revolutions* (1970 [1962]), for example, Kuhn observes that the process of acquiring and transmitting a scientific paradigm through disciplinary methods such as required textbooks, strictly regulated and monitored laboratory research, and so forth, involves "a narrow and rigid education, probably more so than any except perhaps in orthodox theology" (Kuhn 1970 [1962], p. 166). Kuhn describes what he calls "normal science" in terms of a "disciplinary matrix," where "disciplinary" refers both to a scientist's field of specialized knowledge and to the disciplinary and (self-)censoring structure of scientific practice itself. Individual members of a particular scientific community are hardly expected to deviate from the disciplinary matrix–paradigm adopted by that community and are often punished if they do so: they lose credibility, are denied research grants, and may even be tabooed and scapegoated in true tribal fashion.

Kuhn himself remarks that his descriptions of modern scientific practices will inevitably suggest to his reader that a "member of a mature scientific community is, like the typical character of Orwell's *Nineteen Eighty-Four*, the victim of a history rewritten by the powers that be" (ibid., p. 167). Scientists are particularly blind to and will not consider the scientific data that may invalidate and, therefore, threaten their disciplinary paradigm. But, Kuhn, probably mindful of his own precarious, ambiguous position in the scientific community as a philosopher and a historian of science, takes back these observations, when he argues that scientists form a peculiar kind of democratic institution within the larger lay democracy of the nonscientists. And yet, ironically, this argument only reinforces the notion of science as a theological community.

In Wilson's better confessional moments, he does provide a candid inside view of mainstream science, confirming Kuhn's descriptions. For example, he writes: "The most productive scientists, installed in million-dollar laboratories, have no time to think about the big picture and see little profit in it. The rosette of the United States National Academy of Sciences, which the two thousand elected members wear on their lapels as a mark of achievement, contains a center of scientific gold surrounded by the purple of natural philosophy. The eyes of most leading scientists, alas, are fixed on the gold" (Wilson 1998, p. 41).

It is also for such political and lucrative reasons that most mainstream scientific treatises addressed to lay audiences are pretty much interchangeable in terms of the ideas promoted. Predictably, these ideas are the ones that belong to the scientific paradigm that happens to prevail at a given historical moment and that would perpetuate itself at all costs, precisely because of the high professional and financial stakes involved. Even radical dissent is couched in mainstream scientific or political pieties, such as democracy and, most recently, environmentalism. In this respect, Wilson's last chapter, "To What End?" is particularly distressing, even though it might at first blush appear promising, especially by comparison to the rest of the book. Wilson contends that because humanity is a product of millions of years of evolution, it cannot separate itself from the natural habitats in which it has developed without endangering its own future and that of all other life forms. He pleads for a wise management of the Earth, under the guidance of the environmental sciences, and for an overall "existential conservatism" (ibid., p. 332).

Although Wilson makes an eloquent appeal to Western societies to abandon what he calls our "exemptionalist" attitude toward the environment, unfortunately he is in bad faith (in the Sartrean sense): he fails to see that his own absolutist view of Western reductive science is such an exemptionalist attitude. For example, he promotes the theory of the survival of the fittest in the realm of ideas as well, but only if these ideas are taken up and sanctioned by mainstream science. Scientific reductionism itself is largely exempt from such competition. But even when it does engage in a contest, say, with religion or with postmodernist theory, the evolutionary dice are loaded in its favor, so that it is certain to win.

In *The Ecology of Commerce*, Paul Hawken notes that most global problems cannot be solved globally, because "they are global symptoms of local problems with roots in reductionist thinking that goes back to the scientific revolution and the beginnings of industrialism" (Hawken 1993, p. 201). He then proceeds to praise Wilson for apparently inventing the term "exemptionalism" (ibid., p. 204). But Hawken seems unaware that he is praising an exemptionalist wolf in environmental sheep's clothing. In effect, Wilson tacitly exempts a large section of Western mainstream scientists, if not all of them, from human responsibility, presumably because of their professional ethics. According to these ethics, scientists are impartial and disinterested pursuers of objective truth. Therefore, they should by definition be exempt from any social and personal responsibility for the nature of research they engage in and its potentially negative effects on other human beings and the biosphere.

In turn, Wilson lays all of the responsibility at the door of society at large. For example, he writes: "Scientists have been charged with conquering cancer, genetic disease, and viral infection ... and they are massively funded to accomplish these tasks. They know roughly the way to reach the goals demanded by the public, and they will not fail. Science, like art, and as always through history, follows patronage" (Wilson 1998, p. 101). Therefore, Wilson implies, the main function of mainstream science is to put its myriad army of professional scientists to work on whatever their patrons, whether big corporations, governments, or the general public, commission them to research, as long as billions of grant dollars are pumped into this research.

Far from pointing out, like Hawken and others, that reductionist scientific thinking and practices are at the root of some of our most acute local and worldwide problems, Wilson, on the contrary, hails reductionism as the global solution. Wilson does mention the widespread concern that "a science driven society risks upsetting the natural order of the world set in place by God or, if you prefer, by billions of years of evolution" (ibid., p. 36). He also notes that a science that is given too much cultural authority "risks conversion into a self-destroying impiety" (ibid.). Yet, he exempts sociobiology, reductive science at large, and himself from this grim picture.

Wilson stops short of showing, courageously and unequivocally, how reductionist, biological technoscience, for example, is equally responsible for this exemptionalist attitude, creating a destructive feedback loop with the multibillion-dollar, transnational biotechnological and pharmaceutical industry; how a large number of mainstream scientists work busily at providing scientific evidence against the disastrous planetary effects of global warming, polluting chemical agents, industrial waste, and so on; and how a larger number of such scientists are co-opted by the technocratic establishment that confers on them the exalted and well-remunerated status of scientific advisors or "experts." The basic theoretical assumptions and research programs of sociobiology and other reductive science, as outlined in *Consilience*, will unfortunately not help us correct this situation, not least because of the exemptionalist, hubristic mentality of their proponents. Unless they undergo extensive transformation, they will continue leading us down the path of ecological and human regression.

2 Dreams of Global Hegemony: Steven Weinberg's Final Theory

I have examined in some detail the possible adverse impact of the socio-biological notion of consilience within a global reference frame, because, despite wide-ranging, cogent critiques on the part of holistic scientists and philosophers, reductionism remains well entrenched not only among rank-and-file scientists, but also among many a prominent savant other than Wilson. For example, Steven Weinberg, Nobel laureate in physics, in his popular science book on *Dreams of a Final Theory: The Scientist's Search for the Ultimate Laws of Nature* (1992), shares Wilson's view of the global hegemonic mission of Western science, marching under the banner of reductionism. Although Weinberg expresses his scientific fundamentalism in a less radical form than Wilson, he shares the latter's religious commitment to reductionism. After all, his main field of research is elementary particle physics, and he feels compelled to justify and protect massive research programs that amount to billions of dollars in public and private appropriations.

To take a brief example, in the chapter on "Two Cheers for Reductionism," Weinberg reports his exchanges with his "good friend" Ernst Mayr, a holistic, evolutionary biologist who takes him to task for being "an uncompromising reductionist." Weinberg debonairly counters that he is "a compromising reductionist" (Weinberg 1992, p. 53). He then goes on to review various forms of this "heresy" (ibid.). According to him, Mayr identifies three types of reductionism: constitutive or ontological, which is a "method of studying objects by inquiring into their basic components"; explanatory reductionism, for which "the mere knowledge of its ultimate components would be sufficient to explain a complex system"; and theoretical reductionism, which is "the explanation of a whole theory in terms of a more inclusive theory" (ibid., p. 54).

In the end, however, Weinberg claims that he is not speaking, like Mayr, about scientific activities or theories, but "about nature itself" (ibid.). He draws his own distinction between reductionism as "a general prescription for progress in science"—a view he does not share—and reductionism as "a statement of the order of nature," which he believes "is simply true" (ibid.). Moreover, this true order of nature tends to manifest itself at a deeper level in physics than in any other field. Physicists might not be able to explain, say, the properties of DNA molecules in terms of quantum mechanics, so that chemistry "survives to deal with such problems with its own language and concepts." Nevertheless, there are "no autonomous principles of chemistry that are simply independent truths, not resting on deeper principles of physics" (ibid.).

The foregoing citations indicate that Weinberg's views would mainly fall under Mayr's category of theory reductionism—as Weinberg partly concedes—because, like Wilson, he would reduce all theories to the "deeper principles of physics." (Of course, Wilson is guilty of all three reductionist heresies, to use Weinberg's religious vocabulary.) The deeper physical principles would in turn be explained by what some contemporary mathematicians and physicists call a "theory of everything" and what Weinberg calls a "final theory." This theory would unify the strong, electroweak, and gravitational interactions of physical forces in one coherent mathematical formula.

It is equally clear that no genuine dialogue was possible between Weinberg and Mayr to begin with, because the belief in the deeper laws of nature and in the possibility of a final theory comprising one mathematical equation involves a profession of faith that has little to do with rational argument. To be fair to Weinberg, the same thing can be said about Mayr's holistic position. As Weinberg remarks, even though he and Mayr "are still on good terms," they have "given up on *converting* each other" (ibid.; my emphasis). Each would rather preach to his congregation, or to those already converted to his own scientific faith.

Although rational argument cannot usually solve disputes based on faith (as Weinberg himself is aware and as we know only too well from intercultural, religious, and other conflicts), one could nevertheless look at the practical consequences of different views and assess their sociocultural impact in both a local and a global framework. In this regard, Weinberg's professional ethics are as stark, hubristic, and exemptionalist as those of Wilson. He acknowledges the "chilling impersonality" of the laws of nature as conceived by scientific materialism and reductionism (ibid., p. 245). Yet he dismisses the possibility that this chilling impersonality might derive not from nature itself, but from his approach to it. Weinberg's assumption, like that of many other mainstream reductionists, is that humans have a part in discovering the objective laws of nature, but have no part or responsibility in constructing such "laws," which are completely independent of human subjectivity.

Like Wilson, Weinberg is nostalgic for a world that has God at its center and is modeled on a divine, cosmic master plan (ibid., p. 256). He nevertheless estimates that science has relentlessly demystified this age-old belief and will continue to do so all the way to the "final theory." For him, just as for Wilson, science and religion remain engaged in a mortal combat for the minds and hearts of humanity, so that no meaningful reconciliation between them is possible. For this reason, he prefers religious and scientific fundamentalists to their liberal counterparts, who to him appear very much like deserters from their respective camps: "Religious liberals are in one sense even farther in spirit [than scientific reductionists are] from fundamentalists and other religious conservatives. At least the conservatives like the scientists tell you that they believe in what they believe because it is true, rather than because it makes them good or happy" (ibid., p. 257).

The quest for objective truth, in this case for the final theory, is what determines Weinberg's hierarchy of values, life choices, and priorities for planning and funding scientific research. His own field of elementary particles turns out to be the one deserving the highest amount of research dollars, because it comes closest to the final theory. A considerable portion of Weinberg's book revolves around the question of how one could justify the cost of building a Superconducting Super Collider in Texas, to the tune of about eight billion dollars spread over a decade. For Weinberg, the most serious issue is opposition to reductionism that comes not from organismic biologists such as his friend Mayr, but from his own congregation of physicists. According to him, the "reductionist claims of elementary particle physics are deeply annoying to some physicists who work in other fields, such as condensed matter physics, and who feel themselves in competition for funds with the elementary particle physicists" (ibid., p. 54). He shares the view that the proposed building of the Super Collider "is perhaps the most divisive issue ever to confront the physics community" (ibid.). In other words, this issue threatens the tacit pact of physicists to maintain a common front against lay outsiders (*pace* Wilson).

In the prologue of his book, just after confessing that the debate over the Super Collider has forced him into explaining "what we are trying to accomplish in our studies of elementary particles," Weinberg compares himself to Faustus: "For myself, the pleasure of the work had always provided justification enough for doing it. Sitting at my desk or at some café table, I manipulate mathematical expressions and feel like Faust playing with his pentagrams before Mephistopheles arrives" (ibid., p. 5). His Faustian analogy is even more appropriate than Weinberg probably realizes. Could it be that the Super Collider is a Mephistophelian temptation that Weinberg–Faust only too willingly embraces? And are not the dreams of a final theory—which Weinberg mentions, immediately after the Faust analogy, as being what he is really after—even a greater temptation that puts his soul in fatal danger? His Faustus is certainly not Goethe's, but Marlowe's, for in his book there is no final redemption through love and generous service to humanity. What we mostly find are concerns about status, money, and power, which ultimately hurl Marlowe's Faustus into the depths of an agonistic hell.[5]

Weinberg unwittingly expresses the same hubristic attitude through another heroic metaphor, this time borrowed from a Teutonic saga, comparing the early successes of elementary particle physics with Siegfried's violent exploits: "Often we have felt as did Siegfried after he tasted the dragon's blood, when he found to his surprise that he could understand the language of birds" (ibid., p. 4). One can only pray that Weinberg will not be, like Siegfried, betrayed and stabbed in the back, presumably by one of his covetous fellow physicists (of the common, nonelementary particle variety), before he can complete his heroic quest for the final theory. Be that as it may, Weinberg's prodigious scientific talents and not unskillful pen could, no less than Wilson's, be better employed than in providing further justification for an exemptionalist mentality that has transformed much of contemporary Western science into an all-too-willing servant of our global technocratic subculture.

3 General Systems Theory and the Web of Life

The cases of Wilson and Weinberg illustrate that uncompromising, or even compromising, reductionism might not be the best scientific *Weltanschauung* and methodology to further human development in the present global circumstance. One alternative, consilient view that could perhaps better serve this purpose (with appropriate adjustments and transformations) is a certain strand of general systems theory, coupled with cybernetics. General systems theory has its origin in biology, but in an organismic, rather than a reductionist, version of it, also embraced, as we have seen, by Weinberg's friend, Ernst Mayr. In the twentieth century, it was independently developed by Alexander Bogdanov in Russia, during the 1910s, and by Ludwig von Bertalanffy in Austria, during the 1920s.[6]

In turn, cybernetics was developed two decades later by Norbert Wiener who defines it, in the subtitle of his magnum opus, *Cybernetics*, as "the science of control and communication in the animal and the machine" (Wiener 1948). The term comes from the Greek *kybernetes*, steersman, and was used for the first time by French electrodynamicist André-Marie Ampère, in the first half of the nineteenth century. Unlike general systems theory, cybernetics originated not in biology, but in electrodynamics, mathematics, neuroscience, and electronic engineering, ultimately leading to the current explosion of information and communication theory. As employed by Wiener, its main connotations were originally mechanistic, rather than organismic, concerning such concepts as feedback and self-regulation in machines and other closed systems. Cybernetics and general systems theory eventually merged, to the benefit of both fields, through the cross-fertilization and refinement of their main concepts such as open holistic systems, nonlinear causality, self-organization, self-regulation, feedback loops, and so forth.

Such prominent general systems theorists and cyberneticists as von Bertalanffy, John von Neumann, Norbert Wiener, Gregory Bateson, Magoroh Maruyama, and Ervin Laszlo challenge the reductionist model as being a mechanistic, linear, causal paradigm. They define a system not as a static object, but as a dynamic flow of interactions or pattern of events, whose existence and behavior derive not from the nature of its components, but from their manner of organization. A system therefore is more than the sum of its parts, that is, it is nonsummative and irreducible. It maintains and organizes itself by exchanging matter, energy, and information with the environment, that is, with other systems. Thus, systems frame and are framed by other systems in a natural, hierarchical order. According to systems theorists, however, the term "hierarchical" should not be understood in linear fashion, as disciplinary order, but only as larger or more inclusive levels of organization. Systems, subsystems, and suprasystems are nested within one another, rather like sets of Chinese boxes. But, unlike such boxes, they constantly communicate and interact.

By the same token, systems thinkers conceive of the interaction between the organism and the environment not in linear fashion, as reductionist scientists such as Wilson do, but in terms of mutual causality, or causal reciprocity. Unidirectional causal thinking presupposes an essentialist dichotomy between subject and object that depends on static notions of identity, substance, and attributes (Maruyama 1974). By contrast, mutual causality is two-directional, presupposing a dual relation of interdependence and reciprocity between cause and effect.[7] It implies viewing reality as a web of dynamic relations, where substances and identities continually shape and reshape each other and where subject and object are dissolved in a ceaseless play of forces.

Maruyama points out that the nonlinear view of reality has characterized much of human thinking in other parts of the world and throughout history. Consequently, Western mainstream science should equally reconsider its infertile, linear views and fall in step with the rest of the world.[8] To some extent, this has also been my argument throughout part I above, when I challenged the linear views of globalization offered by Western-style, mainstream social science, proposing, instead, a multilevel approach to globality and locality, based in part on the notions of self-organizing systems, emergence, and mutual causality.

Likewise, in defining the concept of global intelligence, I took into account the general systems view of the causal reciprocity between subject and object, mind and body, knower and known. This causal reciprocity involves a much more sophisticated ontoepistemology than Wilson's crude, materialist one. Referring to the relationship between mind and reality, Wilson asserts: "Outside our heads there is freestanding reality. Only madmen and a scattering of constructivist philosophers doubt its existence. Inside our heads is a reconstitution of reality based on sensory input and the self-assembly of concepts" (Wilson 1998, p. 65). He means this assertion as a blow against constructivism, solipsism, and other idealist heresies, without entertaining the possibility of a conceptual alternative to the binary oppositions of matter versus spirit, mind versus body, or materialism versus idealism.

By contrast, Ervin Laszlo, for instance, emphasizes the necessity of transcending precisely such Cartesian dualism in the natural sciences and beyond. For example, he argues that, in approaching experience, we should do away with the dichotomy of subject and object. This does not mean, however, rejecting altogether the concepts of organism and environment, as they have been handed down by natural science. It only means that "we conceive of experience as linking organism and environment in a continuous chain of events, from which we cannot, without arbitrariness, abstract an entity called 'organism' and another called 'environment.' The organism is continuous with its environment, and its experience refers to a series of transactions constituting the organism–environment continuum" (Laszlo 1969, p. 21).

On the other hand, the brain that the neurologist observes cannot be equated, as Wilson and other reductionists do, with the experience of mind to which its synaptic activity corresponds, because brain and mind are not on the same continuum. For this reason, Laszlo develops the concept of "biperspectivism," based on Niels Bohr's principle of complementarity from the new physics, in order to provide an adequate description of the correlation between mind and matter. He means the term to convey both the irreducibility and the complementarity of the mental and the physical worlds: "Whether a system is physical or mental depends on the viewpoint of observation. The operation of passing from the one to the other viewpoint permits the alternate inspection of the complementary (but not simultaneously appearing) aspects" (Laszlo 1973, p. 171).

For Laszlo, matter and mind, body and consciousness are "not ultimate realities," however. They are simply different conceptualizations to bring order into experience, and have "no rigid metaphysical boundaries" (ibid., p. 43). As a conceptualization or metaphor, biperspectivism allows us to deal with both sides of experience without essentializing either of them, as Wilson, Weinberg, and other reductionists do. Laszlo suggests that complex open systems such as human organisms have an exterior and interior dimension, so that mind is coextensive with the physical universe (ibid., p. 293).

The preceding insights are equally helpful in developing a nonlinear concept of global intelligence, as I have attempted to do in the introduction and part I of the present study. This concept privileges neither the material nor the spiritual, neither individual nor collective consciousness, neither human agency nor physical nature or environment, neither the global nor the local. Instead, it regards all of them as complementary parts in an experiential continuum, ceaselessly engaging in causally reciprocal, feedback loops.

In turn, the notion of feedback loops can be no less valuable in approaching globalization, as we have seen in part I above. The concept of feedback loops was first used in cybernetics to describe feedback mechanisms in various closed systems or machines.[9] General systems theory then emancipated it from its mechanistic implications, applying it to open, self-organizing systems, including individual and social organisms. Indeed, causally reciprocal loops can be seen as operating across a large number of fields, from the natural to the social sciences to the humanities. They are also present in evolutionary and environmental theory, where they can account for evolutionary complexity and change in more elegant and plausible fashion than Wilson's and Weinberg's unsophisticated notions of linear causality.[10]

I shall return to the notion of feedback loops, in chapters 4 and 5 below, reinscribing it within a global reference frame and reorienting it toward global intelligence. Here, I would like to note only that the interplay of negative and positive feedbacks offers a plausible account of the complex relationship between the human individual, or social organism, and its environment. Most important, it needs to be taken into account by any theory of global learning, designed to stimulate further human development.

Negative feedback, or what Maruyama (in Milsum 1968) and Laszlo (1973) call Cybernetics I, is a self-stabilizing activity that allows individuals to live in a world that they have constructed through their past experiences. By negative feedback loops, individuals generate conditions in their environment that confirm and correspond to already existing cognitive patterns, through which they in turn perceive this environment. Thus, negative feedback is a form of existential projection. The system, whether it is an individual or a community, projects on its environment the constructs that best match its perceptions.

Changing conditions, however, can lead to a mismatch between a system's perceptions and its constructs, triggering positive feedback loops, or Cybernetics II (Maruyama 1963; Laszlo 1973). Since its experience no longer conforms to its preconceptions, the system develops new constructs in order to change and refine its previous map of the world. This remapping of knowledge is what we call "learning." Just as a natural biological system self-organizes and stabilizes itself in order to adapt to changing conditions, so does a cognitive system, in order to make sense of its new world. According to Laszlo, the feedback process by which the cognitive system generates meaning both alters the environment, through the system's projection of its constructs, and modifies the system itself, as it remaps this environment (Laszlo 1973, pp. 128–131).

For general systems theory, therefore, learning means not merely becoming acquainted with the characteristics and the pattern of organization of an already existing system. Instead, it involves a fundamental reorganization of the system, in which new assemblies occur, different feedback loops arise, and alternate pathways emerge. In the end, both the world and its explorer become transformed through learning. It is in this sense that I am using, in the present book, the notion of local–global learning environments, through which we would learn how to reorient ourselves toward global intelligence. As I have repeatedly pointed out and will continue to do so in subsequent chapters, this learning process will involve an extensive reorganization of all living systems on this planet, including ourselves, and can hardly be achieved overnight.

The notion of positive feedback as learning also implies recognition of the potential creative role of cognitive crises. When obsolete modes of interpretation become dysfunctional, confusion and disorientation set in, because nothing seems to work any more "as it used to"—a complaint that is nowadays heard only too often from all kinds of mainstream practitioners and "experts," called on to puzzle out the emerging global economies and geopolitical configurations. This disorientation can then trigger a generalized state of anxiety and distress, as indeed many of us are experiencing in today's world. Yet, such a state may in turn motivate the system to achieve a more complex level of organization, by seeking and integrating relevant data of which it had previously been unaware or had deliberately ignored. As I have noted in the introduction, it is for this reason that the present global circumstance, with its increasingly acute awareness that things simply cannot go on as before (without a major breakdown in all of our life-support systems) should be seen as an opportunity for, rather than an obstacle to, radical transformation.

For systems thinkers, the feedback processes are constant features observed throughout the cosmos. One can equally perceive them in organic, suborganic, and supraorganic worlds, from atoms to social groups to the biosphere of our planet. Thus, general systems theory allows for the emergence of genuine consilience, quite different from the one Wilson proposes. In such fields as anthropology, biology, information and communication, economics, environmental sciences, humanities, neuroscience, psychology, physics, and sociology, open systems of a biological, electronic, social, psychological, and hermeneutic nature can be treated nonreductively in terms of nonlinear concepts such as dynamic wholeness, mutual causality, feedback loops, self-stabilization and differentiation, information flow, and transformation. Consequently, the nonlinear models pioneered by general systems theory and cybernetics are further developed and refined in many of these branches of learning and could, in turn, be profitably employed in global and intercultural studies, oriented toward global intelligence.

In the life sciences, contemporary developments and refinements include nonlinear concepts such as Ilya Prigogine's "dissipative structures," Humberto Maturana and Francisco Varela's "autopoiesis," Stuart Kauffman's "edge of chaos," Per Bak's "self-organized criticality," Niles Eldredge and Stephen Jay Gould's "punctuated equilibrium," James Lovelock and Lynn Margulis's "Gaia hypothesis," Benoît Mandelbrot's "fractal geometry" and, generally, the "mathematics of complexity" that is associated with dynamic systems theory and is based on nonlinear equations. Together they form a body of scientific thought called chaos and complexity theory (or "chaoplexity") that may lead, in Fritjof Capra's words, to a "unified theory of living systems" (Capra 1997, p. 154f).[11]

Among these conceptual developments, the most relevant to the concerns of the present study is the Gaia hypothesis, which looks at our entire planet as a very complex, self-organizing system. James Lovelock first proposed this hypothesis in 1979, and then developed it throughout the 1980s, together with Lynn Margulis. In his words, the Gaia hypothesis should be considered "as an alternative to the conventional wisdom that sees the Earth as a dead planet made of inanimate rocks, ocean, and atmosphere, and merely inhabited by life. Consider it as a real system, comprising all of life and all of its environment tightly coupled so as to form a self-regulating entity" (Lovelock 1988, p. 12).

The idea of self-regulating systems comes from general systems theory/cybernetics, but the Gaia hypothesis uses it innovatively, by linking together, through a large number of feedback loops, both living and nonliving systems such as plants, animals, humans, rocks, oceans, and atmosphere. For Lovelock and Margulis, Earth is an immensely complex, self-organizing system in which life generates the conditions of its own possibility against almost impossible odds (at least in terms of the second law of thermodynamics).

As Margulis in turn presents it, the Gaia hypothesis assumes that the surface of the Earth, which we have always regarded as the environment of life, is in fact an integral part of life. Consequently, the troposphere (the blanket of air around the earth) should be regarded "as a circulatory system, produced and sustained by life" (Margulis 1989). According to her, linear and reductionist, evolutionary scientists (such as Wilson) "perpetuate a severely distorted view" when they claim that life adapts to an inanimate environment of chemistry, physics, and rocks. Quite the contrary, life creates and changes the environment to which it adapts. In turn, that environment "feeds back on the life that is changing and acting and growing in it. There are constant cyclical interactions" (Margulis 1989). Evolution cannot be limited to adaptation or to a static environment. As Lovelock puts it, "So closely coupled is the evolution of living organisms with the evolution of their environment that together they constitute a single evolutionary process" (Lovelock 1988, p. 99).

Most important for an emergent ethics of global intelligence, the Gaia hypothesis shifts the focus of scientific research not only from evolution to coevolution, but also from adaptation, survival of the fittest, and random variation to mutual dependence, creativity, and cooperation. In other words, it assigns a central evolutionary role to symbiotic processes. Capra, commenting on the theory of symbiogenesis, a hypothesis first elaborated by Konstantin S. Merezhkovsky (1909) and then, over half a century later, by Margulis, points out that all "larger organisms, including ourselves, are living testimonies to the fact that destructive practices do not work in the long run. In the end, the aggressors always destroy themselves, making way for others who know how to cooperate and get along. Life is much less a competitive struggle for survival than a triumph of cooperation and creativity. Indeed, since the creation of the first nucleated cells, evolution has proceeded through ever more intricate arrangements of cooperation and coevolution" (Capra 1997, p. 238).

Starting from the Gaia hypothesis, Capra proposes the metaphor of the "web of life" for describing Earth as a vast network of interacting and mutually supportive self-organizing systems. This metaphor is grounded in mutual dependence and cooperation at all biological and cultural levels, rather than in a raw, Wilsonian struggle for life and survival of the fittest. As Margulis and Sagan put it, "Life did not take over the globe by combat, but by networking" (Margulis and Sagan 1986, p. 17). Moreover, in the systems view that Capra shares with the proponents of the Gaia hypothesis, life is far from limiting itself to survival and reproduction. Rather, it exhibits an "inherent tendency to create novelty, which may or may not be accompanied by adaptation to changing environmental conditions" (Capra 1997, p. 221).

The web of life also implies a scientific notion of order and disorder that is entirely different from that of sociobiologists such as Wilson or of social scientists such as Jowitt, Appadurai, and Bauman. In the traditional scientific view shared by the last four scholars, order is associated with equilibrium, usually in static structures, and disorder, with nonequilibrium situations such as turbulence. By contrast, according to chaos and complexity theory, nonequilibrium or "a state far from equilibrium" is a source of order. As Capra remarks, the "turbulent flows of water and air, while appearing chaotic, are really highly organized, exhibiting complex patterns of vortices dividing and subdividing again and again at smaller and smaller scales. In living systems, the order arising from nonequilibrium is far more evident, being manifest in the richness, diversity, and beauty all around us. Throughout the living world, chaos is transformed into order" (ibid., p. 185). Whereas for mainstream science, chaos/disorder is the negative term of a conflictive binary opposition, for the chaos and complexity theorists it is an integral part of a dynamic correlation based on mutual or circular causality.

Now it should be easy to see why a nonlinear evolutionary model such as the "web of life" would be more appropriate for learning and research in the life sciences within a global framework than Wilson's consilience and other reductionist models. Whereas the latter models perpetuate the globalitarian pretensions of Western science and other institutions of an earlier age, the former encourages and supports a cooperative, symbiotic view of evolution, in which all living and nonliving components of the overall global system depend on each other for their well-being and sustainable development. The nonlinear evolutionary model of the web of life also stresses creativity and diversity, rather than uniformity, as key factors in both natural and cultural evolution. Most important of all, it moves away from exemptionalism by stressing the great responsibility that each self-aware member of the global system has toward other such members, as well as toward life in general.

Finally, the web of life is an idea that certainly does not belong exclusively to Western culture. It can equally be found, in one form or another, in all the major cultural traditions of the world. Consequently, it would constitute an excellent point of departure for global, intercultural research and dialogue. In the next two chapters, I shall attempt to lay the foundations for such an intercultural dialogue that I hope will eventually lead to an intercultural ecology of science, oriented toward global intelligence.

4

Buddhist, Taoist, and Sufi Views of the Web of Life: An Intercultural Comparison

In the previous chapter, we have seen that the principles that inform the Western nonlinear models of self-organization and evolutionary theory could be more productive and helpful than those of scientific reductionism in the current global circumstance. But they would also need to undergo thorough self-examination and appropriate adjustments, if they are to provide a fruitful basis for extensive intercultural scientific research, dialogue, and practice. To this purpose, it would be useful to compare them with a number of nonlinear views of life that are present outside Western civilization, notably early Buddhist and Taoist thinking, as well as later, Islamic Sufism.[1]

Joanna Macy has already carried out some of this research, concentrating on the nonlinear views present in early Buddhist thinking.[2] Moreover, she proposes a mutual hermeneutic between early Buddhism and general systems theory: she interprets each conceptual system in terms of the other in such a way that both of them emerge modified, amplified, and enriched. Through this mutual or reciprocal hermeneutic, Macy hopes to build what she calls a "Dharma of natural systems," which would in turn constitute the "philosophic basis and moral grounding for the ecological worldview emerging in our era" (Macy 1991, p. xii). In the wake of her friend and collaborator, the Norwegian thinker Arne Naess (1973, 1989), she calls this worldview "deep ecology" and describes it as "symbiotic, synergistic, pluralistic, and mutualistic" (ibid., p. 17). Needless to say, I subscribe, in part, to this intercultural, deep ecology. I also believe, however, that some of its basic assumptions need to be revised. Therefore, in the present chapter and the next one, I shall attempt not only to enrich it with enlightened Taoist and Sufi voices, but also to reorient it toward an emergent ethics of global intelligence, grounded in a mentality of peace.

1 Mutual Causality or Dependent Origination in Early Buddhist Thought

Macy convincingly demonstrates that reciprocal causality is a cornerstone of Gautama Buddha's thought and practice, as revealed in his teaching of *paṭtica samuppāda*, which can be translated as "dependent co-arising" or "dependent origination." In marked contrast to the previous essentialist and substantialist doctrines of the Vedic tradition, which were based on perceptions of unidirectional, linear causality, early Buddhist thought shifts to "perceptions of dynamic interdependence where phenomena affect each other in a reciprocal or mutual fashion" (Macy 1991, p. 1).

As Macy points out, the Buddha changes the traditional, Vedic definition of causality to express dynamic relationships rather than substance. For instance, in part II of the *Saṃyutta Nikāya*, called *Nidānavagga* or the Book of Causation, the Buddha explores the *nidānas* or "causes" of suffering, such as ignorance, volitional formations, consciousness, name-and-form, and so on, by asking the monks (bhikkhus): "And what, bhikkhus, is dependent origination? With ignorance as condition, volitional formations [come to be]; with volitional formations as conditions, consciousness [comes to be]; with consciousness as condition, name-and-form; with name-and-form as condition, the six sense bases; with the six sense-bases as condition, contact; with contact as condition, feeling; with feeling as condition, craving; with craving as condition, clinging; with clinging as condition, existence; with existence as condition, birth; with birth as condition, aging-and-death, sorrow, lamentation, pain, displeasure, and despair come to be. Such is the orgin of this whole mass of suffering. This, bhikkhus, is called dependent origination" (*Saṃyutta Nikāya*, II.1).

The Buddha is not interested in finding out what "causes" produce a given factor *A* (such as suffering, ignorance, or craving). Rather, he seeks to determine what else happens in relation to the happening of *A*. In this sense, the occurence of *A*, say, craving, provides a locus or context in which *B*, ignorance, can equally occur. Or, put in another way, *B* arises codependently with *A*. So ignorance is not the "prime cause" of suffering as later Buddhist scholars assumed. Rather, it is present with the other *nidānas* when suffering occurs.

But Macy also cautions against understanding the relation of *A* and *B* simply as a contiguity of events, as in David Hume's view, which Western scholars have occasionally mistaken for the Buddhist notion of causality (Macy 1991, p. 53). For Hume, physical events flow past us, but are objectively unrelated, and it is only our mental operations that infer a causal connection between them. By contrast, the Buddha perceives both an ontological and an epistemological connection between events, so that mutual causality is the way things happen or the "nature of things" (*dhammatā*).

Macy further points out that earlier Buddhists did not attempt a "metaphysical analysis in terms of discrete entities," or dharmas (ibid., p. 60). Early Buddhism is not an analytic but a synthetic view, "involving an awareness of wholeness—a wide and intent openness or attentiveness wherein all factors can be included, their interrelationships beheld" (ibid., p. 63). It was only later on, in the *Abhidharma Piṭaka*, that scholars elaborated the philosophical aspects of the Buddha's teachings, interpreting the dharmas as discrete entities, or "the fundamental building blocks into which conventional reality can be dissected" (ibid., p. 59). Nor does the earlier Pali canon describe emancipation in terms of an escape from causality, as the Abhidharmists and subsequent substantialist interpreters do. On the contrary, emancipation can be reached only by understanding dependent origination or reciprocal causation and, thus, "using the leverage of conditionality" (ibid., p. 60).

In the *Abhidharma*, on the other hand, there is "a shift toward a more substantialist and linear view, where effects preexist in their causes and are produced by them" (ibid., p. 61). Furthermore, the *Abhidharma* introduces a categorical distinction between the mental and physical realms, between reality and appearance. Not unlike the later Chinese legalist scholars (mentioned by Needham and then Wilson) who interpreted Taoist and Confucian thought in analytic, reductive ways, the Abhidharmist scholars reinterpreted the earlier books of the Pali canon to fit their reductionist and substantialist views. Such views had, moreover, been prevalent in pre-Buddhist, Vedic texts, so that Abhidharmist reductionism was "an unfortunate drift back into essentialist thinking"

(Streng 1975, p. 79). So much, then, for Wilson's Western supremacist claim that reductionism was a lucky invention of the European Age of Enlightenment.

The Buddhist view of subjectivity or human agency is similar to that of general systems theory: both views dissolve the subject/object dichotomy into a reciprocal play of relations between fluid variables. The linear causal paradigm in both Vedic India and the modern West exclusively emphasizes either the perceiver or the perceived in the cognitive process. We have seen, for instance, that empiricists such as Wilson believe that the world is the cause of our perceptions, registering its data on neutral sense organs. The symmetrical opposite view, in philosophical solipsism, is that external phenomena are our own projections. By contrast, the Buddha denies neither the being there of the sense objects nor the projective tendencies of the mind, but regards the two as mutually conditioned: we both shape and are shaped by sensory experience—an insight that, as we have seen, Ervin Laszlo and other general systems theorists equally share.

The nonlinear views of early Buddhism and Western systems theory, according to which everything is codependent and in flow, do not allow us to entertain reductionist dreams of a final theory and of immutable physical laws. Yet, the impossibility of achieving ultimate formulations of reality need not be seen, in the manner of Wilson and Weinberg, as cognitive relativism (where "anything goes") or a defeat of human reason. As Macy remarks, it is only "final assertions that are suspect, not the process of knowing itself. It is the illusion that the knower is separate from and unconditioned by the world she would know that drives her into error and derails her pursuit of truth. When the dependent co-arising nature of her mental processes is acknowledged, then her knowing enhances her conscious connection with and participation in the reality of which she is a part" (Macy 1991, p. 130).

2 Mutual Causality in Early Taoism

The concept of dependent origination or reciprocal causality is equally present in the three major strands of traditional Chinese thought, namely Taoism, Confucianism, and Buddhism (the last one came to China from

India, during the first century A.D.). Taoism is believed to be the most ancient, its founding being attributed to the poet–sage Lao Tzu (604–531 B.C.), an older contemporary of Confucius and Gautama Buddha. Although early Buddhism and Taoism obviously developed separately and in parallel for over five hundred years, they have many points in common, including the notion of the cosmos as process, or as a web of dynamic relations (rather than independent substances or identities).

Lao Tzu's nonlinear way of thinking, moreover, just like that of the Buddha, will also undergo essentialist interpretations later on in the tradition, at the hands of both Eastern and Western scholars. But, most important from the perspective of an emergent ethics of global intelligence, both ways of thinking share a nonviolent, peaceful approach to the human and natural worlds, as well as to the web of life in general. For the purposes of the present argument, I shall briefly look at early Taoism, although both Confucianism and Chinese Buddhism would yield important insights regarding the kind of thinking that might work well in the current global circumstance and should equally become the object of extensive intercultural research teamwork.[3]

The Chinese word "tao" is roughly equivalent to the Sanskrit "dharma" (or the Pali "dhamma") and can, like the latter, be translated into English as "path," or "way," or "nature" (of things). Early Taoism, just like early Buddhism, is a practice, more than a philosophy, being concerned with the most appropriate ways of living, based on a proper understanding of the nature of reality. This reality is seen as process or as continuous flow, where all things rise and fade through the formation and dissolution of interactive networks of relationships, based on reciprocal or cyclical causality. Like early Buddhism, Taoism proposes a way of emancipation for the individual, not so much by escaping the cycle of mutual causality as by realizing its nature and becoming one with it. The enlightened Taoist practitioner, just as his Buddhist counterpart, is able to ride up and down the billow of existence with steady equanimity. To use a modern metaphor, they are not unlike skillful surfers who maintain their balance by moving with—and yet as if above—the surging and falling surf.

Nonlinear causality in Taoism is expressed in terms of a ceaseless play of polarities.[4] Just as in early Buddhism (and in general systems theory), polarities in ancient Chinese thinking are metaphorical and paradoxical ways in which one can describe fluid relationships among elements with ever-changing identities. Ancient Chinese thinking, therefore, no less than its early Buddhist counterpart, "entails an ontology of events, not one of substances" (Hall and Ames 1987, p. 15). In this kind of process ontology, human events and agencies are not understood in terms of qualities, attributes, and substances, but in terms of specific contexts and relations. As such, process ontology "precludes the consideration of either agency or act in isolation from the other. The agent is as much a consequence of his act as the cause" (ibid.).

No less than early Buddhism, then, early Taoism (as well as early Confucianism) makes no substantialist distinction between subject and object, but operates instead with a dynamic process of fluid interrelationships, based on circular causality. It follows that the Tao, like the Dharma, can never be named, not because it does not exist, but because its existence is in constant flow. It can therefore be expressed only in a paradoxical fashion, as the nameless that produces all names, the no-beginning of all beginnings, the Nothing that produces Something, and so on. For example, the beginning verses of *Tao Te Ching* read:

The way that can be spoken of
Is not the constant way;
The name that can be named
Is not the constant name.
The nameless was the beginning of heaven and earth;
The named was the mother of the myriad creatures.
. . .
These two are the same
But diverge in name as they issue forth.
(I.1)

In book I, chapter 2, circular causality is applied to polarities such as beautiful and ugly, good and bad, difficult and easy, high and low, and so forth:

Thus Something and Nothing produce each other;
The difficult and the easy complement each other;
The long and the short offset each other;
The high and the low incline toward each other;
Note and sound harmonize with each other;
Before and after follow each other.

The first two polar elements, "Something" and "Nothing," as well as the last two, "before" and "after" clearly show that the Tao should be conceived not as an essentialist entity/identity that splits itself into the manifold (or engages in a dialectic of one and the many), but as a dynamic process of cyclical causality, where there is no primordial cause or absolute beginning.

The *Chuang Tzu*, the other major work in the Taoist canon, expresses this paradox of circular causality very well, in the form of an aporia: "Being a beginning. Being not yet beginning to be a beginning. Being not yet beginning to be a not yet beginning to be a beginning. Being being. Being nonbeing. Being not yet beginning to be nonbeing. Being not yet beginning to be a not yet beginning to be nonbeing. Then suddenly, being nonbeing. And when it comes to being nonbeing, I don't know yet what's being and what's nonbeing" (Chuang Tzu, *Inner Chapters*, II.15).

This aporetic language should, however, not be mistaken for that of certain strands of Christian negative theology, where a linear, hierarchical, transcendental God produces all reality and eventually dissolves warring opposites into an all-embracing unity; nor should it be understood as a Hegelian dialectics that eventually overcomes all contraries by sublation or higher synthesis. As the line "Before and after follow each other" (in the *Tao Te Ching*) suggests, a circle or a ring is the best way of imagining the Taoist cycle of being and nonbeing, of beginning and not-yet-beginning: any point on a ring is both before and after any other point, depending on the arbitrarily chosen starting place.

Early Buddhism is equally emphatic on the impossibility of determining a primordial or "uncaused" starting point, such as the Aristotelian Unmoved Mover. For example, in the Book of Causation, the Buddha says about samsara or the beginningless round of rebirth: "Bhikkhus, this samsara is without discoverable beginning. A first point is not discerned of beings roaming and wandering on hindered by ignorance and fettered by craving. Suppose, bhikkhus, a man would reduce this great earth to balls of clay the size of jujube kernels and put them down, saying [for each one]: 'This is my father, this my father's father.' The sequence of that man's fathers and grandfathers would not come to an end, yet this great earth would be used up and exhausted. For what reason? Because, bhikkhus, this samsara is without discoverable beginning" (*Saṃyutta Nikāya*, II.179).

3 Irenic Mentality and the Taoist Concept of *Wu Wei*

In Taoism, as in early Buddhism, the idea of dependent co-arising is explained in terms of feedback loops that allow us to adopt the most appropriate behavior, that is, the one that comes closest to the way:

> Turning back is how the way moves;
> Weakness is the means the way employs.
> The myriad creatures in the world are born from Some-thing, and Something from Nothing.
> (*Tao Te Ching*, II.XL)

The verse that proclaims weakness to be a preferred strategy of Tao—and there are many such verses throughout the *Tao Te Ching*—might lead one to regard the Tao as a quietist philosophy, comparable to ancient Stoicism within the Western tradition. On this widespread view, the Tao would simply be a way of negotiating the world of power by devising either a submissive or a quietist code of conduct, unthreatening to the latter. Hence the *wu wei* doctrine, according to which "the sage keeps to the deed that consists in taking no action and practices the teaching that uses no words" (*Tao Te Ching*, I.II).

Wu wei, however, does not imply doing nothing at all. As we can see from the foregoing verse, even nonaction is a "deed" or a form of action. Instead, *wu wei* implies, on the one hand, noninterference with the natural process that is Tao and, on the other hand, engagement in actions that are consonant with it. Thus, the art of *wu wei* consists in both letting nature take its course and acting in accordance with its processes. For instance, one should allow a river to flow toward the sea unimpeded and not erect a dam that would interfere with its natural course. Above all, however, *wu wei* means moving away from competitive or violent deeds, toward generous, supportive, and enabling action:

> Do that which consists in taking no action; pursue that which is not meddlesome; savor that which has no flavor
> Make the small big and the few many; do good to him who has done you an injury.
> (*Tao Te Ching*, II.LXIII)

The injunction to "do good to him who has done you an injury" is of course familiar to the Western reader from the New Testament, which equally stresses the cultivation of meekness and the "turning of the other cheek." In a related passage, the *Tao Te Ching* also advances the notion of the leader as servant, which is, again, very much in the spirit of the New Testament:

Therefore, desiring to rule over the people,
One must in one's words humble oneself before them;
And, desiring to lead the people,
One must, in one's person, follow behind them.
Therefore the sage takes his place over the people yet is no burden; takes his place ahead of the people yet causes no obstruction. That is why the empire supports him joyfully and never tires of doing so.
It is because he does not contend that no one in the empire is in a position to contend with him.
(II.LXVI)

The preceding lines make it clear that one ought to translate *wu wei* as "nonviolent" or "noncompetitive action," if one wishes to understand its proper meaning. One may, of course, say with D. C. Lau and other scholars that the Tao places high value on physical survival, suggesting ways in which one could stay alive in troubled times such as the Age of the Warring States (during which the teaching of the Tao supposedly arose). One way is following the ancient principle of "Bowed down then preserved" (I.XXII). Another one is cultivating the Three Jewels or "Treasures," equally present in Buddhism, namely: moderation or "frugality"; self-effacement or "not daring to take the lead in the empire"; and "compassion" (II.LXVII). Finally, you can withdraw from the power contest altogether, so that those in power may find no fault with you. To this effect, Lao Tzu says:

Highest good is like water. Because water excels in benefiting the myriad creatures without contending with them and settles where none would like to be, it comes close to the way.
In a home it is the site that matters;
In quality of mind it is depth that matters;
In an ally it is benevolence that matters;
In speech it is good faith that matters;
In government it is order that matters;
In affairs it is ability that matters;
In action it is timeliness that matters;
It is because it does not contend that it is never at fault.
(I.VIII)

These quiet practices might appear, at least to a mentality of power, as bowing to authority. Even so, they are effective in defusing violence and solving seemingly intractable conflicts through peaceful means and to the benefit of all parties involved. Therefore, they do not necessarily belong to a pragmatic, survivalist doctrine, according to which the Taoist practitioner would, not unlike certain present-day Darwinists, favor the "weak" or prudent position over the "strong" one, as a strategy for survival. On the contrary, they show that the early Taoist, like the early Buddhist (and like the early Christian, one might add) practices a way of being, thinking, and acting that goes beyond the world of power and well beyond a survivalist mentality. In the terminology that I employed in chapter 1 above, the world described by the Taoist (or the early Buddhist or Christian) is incommensurable, rather than merely incompatible, with the world of power. As such it is entirely alien to his audience—mostly rulers and their "vessels" or imperial functionaries who have never experienced the Tao and cannot even imagine it. Therefore, the Taoist practitioner has little choice but to refer to it in terms of a vocabulary of power.

Yet, as we can see throughout the *Tao Te Ching*, Lao Tzu constantly turns this vocabulary around, problematizing it through aporia and paradox. For example, he claims that the Tao employs weakness, nothingness, or emptiness, and all the other virtues that a mentality of power depreciates as negative or undesirable. By reversing the asymmetrical polarities of good and bad, weak and strong, submissive and assertive, empty and full, bright and dull, crooked and straight, male and female, mountain and valley, and so forth, the Taoist practitioner does not simply restore the balance of power or celebrate the cyclical process of *coincidentia oppositorum*. Rather, he emancipates the "weaker" term from the polar structure itself, showing it to belong to a different system of values and beliefs that does not function in bipolar terms.

So, strictly speaking, early Taoism is not polar thinking after all, even as it employs such thinking for its own purposes. In the vocabulary of systems theory, the Taoist emancipates the "weaker" pole from the feedback cycle in which a power dialectics has locked it and where, in due course, "strong" inevitably becomes "weak," and "weak" becomes "strong." Then, he allows each of these poles to create its separate amplifying feedback loops. In these new loops, the strong or the violent will not become weak, but will simply breed more of the same:

One who assists the ruler of men by means of the way does not intimidate the empire by a show of arms.
This is something which is liable to rebound.
Where troops have encamped
There will brambles grow;
In the wake of a mighty army
Bad harvests follow without fail.
(I.XXX)

War will bring with it more war, sowing destruction rather than bountiful harvests for friends and foes alike. If one must use arms, one should do so without relish and without gloating over a defeated enemy. Victory is as undesirable as defeat, because both are part of the same destructive feedback loop:

Arms are instruments of ill omen, not the instruments of the gentlemen. When one is compelled to use them, it is best to do so without relish. There is no glory in victory, and to glorify it despite this is to exult in the killing of men. . . . When great numbers of people are killed, one should weep over them in sorrow. When victorious in war, one should observe the rites of mourning. (I.XXXI)

The Taoist practitioner does not try to change the world by force, for he knows that force will result in more force. Therefore, he "avoids excess, extravagance and arrogance" (I.XXIX). At the same time, he cultivates meekness, which will in turn generate more meekness. One should treat all people with kindness, indifferent of their nature. In this way, one will amplify the feedback loop of kindness, instead of engaging in a conflictive mimetic relationship and thus relapsing into a destructive feedback cycle:

Those who are good I treat as good. Those who are not good I also treat as good. In so doing I gain in goodness. Those who are of good faith I have faith in. Those who are lacking in good faith I also have faith in. In so doing I gain in good faith. (II.XLIX)

The Tao is a way of being and acting in the world that is based on a transvaluation of power-oriented values, which explains its paradoxical nature in relation to these values. Thus, within a system organized around the power principle, "Truthful words are not beautiful; beautiful words are not truthful. Good words are not persuasive; persuasive words are not good. He who knows has no wide learning; he who has wide learning does not know" (II.LXXXI).

In turn, meanness, contentiousness, and hoarding knowledge are characteristics of the imperial "experts" and their mentality of power. At the same time, care for the other, boundless generosity, and gentleness are the mark of the Taoist sage, who thus also redefines the notion of wealth prevailing in his time (and in ours):

The sage does not hoard.
Having bestowed all he has on others, he has yet more;
Having given all he has to others, he is richer still.
The way of heaven benefits and does not harm; the way of the sage is bountiful and does not contend.
(II.LXXXI)

Likewise, if a student of Tao is not prepared to give up the power-oriented system of values and beliefs he has been socialized into, he will completely miss the Taoist teaching, or will find it unusable, if not laughable:

When the best student hears about the way
He practices it assiduously;
When the average student hears about the way
It seems to him one moment there and gone the next;
When the worst student hears about the way
He laughs out loud.
If he did not laugh
It would be unworthy of being the way.
(II.XLI)

In turn, it is only within the reference frame of power that "Straightforward words/Seem paradoxical" (II.LXXVIII), and the way seems dull, or vague, or empty, or nameless:

The way that is bright seems dull;
The way that leads forward seems to lead backward;
The way that is even seems rough.
The highest virtue is like the valley;
The sheerest whiteness seems sullied;
Ample virtue seems defective;
Vigorous virtue seems indolent;
Plain virtue seems soiled;
. . .
The way conceals itself in being nameless.
(II.XLI)

Even from this brief discussion, it is evident that, in early Taoist and Buddhist thinking, reciprocal causality and other nonlinear concepts co-arise with a life experience that is different from the one correlated with linear ways of thinking. Once we understand the cyclical character of the Dharma or the Tao, we can take appropriate steps to intervene, not against but in consonance with it, in order to transform ourselves, as well as the reality that co-arises with us. The fact that reality will change along with us is perhaps the most valuable insight to be derived from this ancient Eastern wisdom, for which being human represents a great opportunity, as well as an enormous responsibility.

Neither Buddhist nor Taoist thinking posits a discontinuity between human and other forms of life. We are all integral parts of the Dharma or the Tao. For Buddhism, humans are but one manifestation of a psychosomatic, cyclical flux, arising now as god, now as animal, now as plant, now as human. For Taoism, humans are equally part of the returning life flow. For both the Buddhist and the Taoist practitioner, however, humans differ from other forms of life by their ability to make choices through self-awareness. As Macy aptly puts it, only "the human possesses the power to choose and change—hence the rare and priceless privilege of a human life; hard to win, it brings both responsibility and the possibility of enlightenment" (Macy 1991, p. 153).

The ability to choose that is unique to humans should not be seen, however, as an occasion for exemptionalism. On the contrary, humans remain codependent with all other living and nonliving systems and cannot act alone, in disregard of the Dharma or the Tao, without adverse consequences for the larger systems within which they are nested. Hence the increased responsibility of humans who are answerable not only to each other, but also to all other forms of life on Earth.

4 Liminality, Incommensurability, and Alternative Worlds

Early Buddhism and early Taoism also describe the ways in which humans can attain the self-awareness needed to make the most appropriate existential choices. One should, again, keep in mind that Buddhism and Taoism are not so much philosophical or scientific theories as natural and social practices. The Buddha, for example, is interested in describing the organizing patterns, structures, and external correlations of various systems (to adopt a general systems vocabulary) not for the sake of objective knowledge or the development of science, but to show practitioners how to reorganize experience by attending to its processes.

Meditation is one of the techniques by which the practitioner stands back from conceptualization in order to attend to its workings. Conceptualization tends to operate within preestablished cognitive frameworks so that, through meditation, the practitioner learns how to detach himself from such frameworks, avoiding the perpetuation of their validity. In turn, early Taoism prefers the practice of "deconstructing" preestablished concepts through logical paradox in order to reorganize experience in consonance with the Tao—a technique that is also available in early Buddhism, as well as in later schools, such as Ch'an/Zen Buddhism or Islamic Sufism.

Such techniques are enhancements of what systems theory calls "objectless knowing" or "second-order thinking" by which cognitive feedback loops, or the very process of cognition, can in turn be known. It is a metalevel of reflexive consciousness, in which one comes to know the knowing of knowing, as the *Chuang Tzu* would probably phrase it. In an often cited passage from *Udāna*, the Buddha describes it in paradoxical terms that are nearly identical to those in which Lao Tzu describes the Tao: "There is, monks, that sphere (*āyatana*) wherein there is neither earth, nor water, nor fire, nor air; wherein is neither this world nor a world beyond, nor moon, nor sun. There, monks, I declare, is no coming, no going, no stopping, no passing-away and no arising. It is not established, it continues not, it has no object. This is indeed the end of suffering" (*Udāna*, p. 80).

Macy notes that *āyatana*, here translated as "sphere," means literally "gateway." According to her, *āyatana* is "the means by which we perceive, or the way in which we perceive, rather than an objective self-existence, supernatural essence or realm" (Macy 1991, p. 134). Although in this cognitive state the knower transcends the world of the six senses (the sixth sense, in Buddhism, being the faculty that perceives imaginary objects), she "would be very much in touch with what permits her interaction with the environment. She would be conscious of that by which she knows it and, therefore, by which she can 'step out' from the fabrications she imposes on it" (ibid., p. 135). Macy calls this state *nondiscriminatory awareness*.

The notion of nondiscriminatory awareness is equally operative in Taoism, even though Lao Tzu, when mentioning possible ways of attaining such awareness, does not seem to refer specifically to meditation, but to retreat from contest, quiet self-reflection, and deconstruction of polar thinking through paradox. As we have seen in the *Tao Te Ching* I.1, for the Taoist the "nameless" and "the named" are the same, but "diverge in name as they issue forth." The Tao refers to a state of awareness, always present, without beginnings or endings, that transcends the sensory experience by which we make distinctions in our interaction with the environment, apprehending and ordering phenomena. The early Taoist, no less than the early Buddhist, encourages the practitioner, through whatever means, to cultivate a mode of perception that goes beyond objects to what brings them forth in the process of dependent co-arising of perceiver and perceived.

I should add that the Pali term *āyatana*, interpreted as "gateway," would also justify describing nondiscriminatory awareness in terms of a liminal state that allows us not only to know the ways in which we know, but also to shift into other modes of perception or cognitive reference frames, that is, to reorganize experience in more or less radical ways. I have already begun an exploration of the concept of liminality in part I above, emphasizing its potential usefulness for intercultural studies, based on an emergent ethics of global intelligence. Here I would like briefly to explore its role as a passageway between worlds or reference systems that are incommensurable, in order to indicate how a paradigmatic shift in human mentality could actually take place.

In *Aṇguttara Nikāya* V.7, the Buddha seems to refer to a liminal state when he addresses Ananda's question about the virtues of meditation: "Could there be such an attainment of concentration that the monk will not be conscious of earth in earth, not of water in water ... and yet he will be conscious?" The Buddha replies: "Herein, Ananda, a monk is thus conscious: 'this is peace, this is excellent, namely, the calming down of all formations ... destruction of craving, detachment, cessation, nibbana." So, the Buddha is interested in *āyatana* not only as a metafaculty of perception, or of knowing how we know, but also as a means of transcending specific worlds, organized according to the power principle. In this liminal state, the practitioner attains peace or "the calming down of all formations."

But, this calming down or temporary peace does not necessarily guarantee access to alternative worlds that organize themselves on principles other than power. Such worlds remain to be constructed through the correlation and causal reciprocity of the interior and the exterior aspects of phenomena. That is why the Buddha still describes nondiscriminatory awareness in the negative terms of "destruction of craving, detachment, cessation, nibbana." The Pali word *nibbāna* (or the Sanskrit *nirvāna*) signifies "blowing out," as in blowing or snuffing out a candle flame, and therefore is an equally negative description, not unlike the Taoist "Nothing." Even the word "peace" is not yet emancipated from its negative meaning of absence of craving, and so forth. All of these terms belong to the world that the practitioner has left behind, rather than to an alternative world. In this sense, nondiscriminatory awareness can be described as a liminal state or passageway between what has arisen and what is yet to arise.

Chögyam Trungpa Rinpoche (1987) refers to a similar liminal state in his commentary to *The Tibetan Book of the Dead*, when he explores the notion of *bardo* that designates the interval between death and (the next) birth. According to the Rinpoche, the word *bardo* signifies gap, being a compound of *bar*, "in-between" and *do*, "island" or "mark." Hence *bardo* is "a sort of landmark which stands between two things. It is rather like an island in the midst of a lake. The concept of *bardo* is based on the period between sanity and insanity, or the period between confusion and the confusion just about to be transformed into wisdom; and of course it could be said of the experience which stands between death and birth. The past situation has just occurred and the future situation has not yet manifested itself so there is a gap between the two" (Trungpa 1987, pp. 10–11).

Although the Rinpoche refers to a particular transitional state between death and the next birth, this state displays all the features of liminality. *Bardo* is the interstice between two conditions or two worlds. And even though it shares some characteristics of both, it properly belongs to neither. It is fraught with great promise, but also with great danger. For instance, it is in this state, which as a rule spans a forty-day period after death, that the "bewildered" mind or consciousness of the dead Buddhist monk is presented with a wide variety of good or bad choices for the next birth. The indeterminate nature of the *bardo* also points to the fact that liminal worlds are not alternative worlds, because they have no firmly established organizing principles and reference frames. Rather, they are indeterminate ontological landscapes, where a great number of such principles and reference frames may arise and present themselves to consciousness for consideration and final choice.

But, the monk is being assisted by his fellow monks even in the bardo state. They continue to read to him from the book, instructing him as to what to expect next and reassuring him through some of the frightening experiences that arise in this state. Even so, the right choice for the next state or world depends solely on the deceased monk. Finally, for the Buddhist practitioner death itself is a liminal state or passageway between two worlds and therefore needs to be seen in its proper perspective. (This is true for many other religious practitioners, including Christians who, if Catholic, have their own version of postmortem liminality, in the concept of purgatory or "limbo.") On the one hand, death does not have the overriding importance accorded to it by a certain (materialist) mentality of power, for example, in the notions of "struggle for life" and "survival." On the other hand, instead of being looked upon as an unavoidable, great misfortune, it ought to be prepared carefully and joyfully as a great opportunity not to be missed.

Specifically, death or the *bardo* can be seen as a liminal space between incommensurable systems or worlds. As I have remarked in connection with early Taoist thinking, power-oriented worlds and those based on a different organizing principle, such as peace, are incommensurable, rather than incompatible. As such they do not clash, but operate with entirely different reference frames. At the same time, however, power attempts to co-opt certain irenic values to serve its own purposes, as it does with any values and/or reference systems that are alien to it—such is power's nature (and modus operandi). In this regard, the general systems that are described by von Bertalanffy, Wiener, Maruyama, Laszlo, Capra, Macy, and others, and to which the Buddha and Lao Tzu equally refer, are power-oriented systems.

The question remains whether one could transit from one incommensurable world to another, specifically from a power-oriented world to an irenic one, and, if so, how? In my view, liminal spaces are precisely interstices between systems and polysystems or worlds that facilitate such transitions (Spariosu 1997, pp. 58–66). The *bardo*, therefore, constitutes the kind of liminal interstice in which humans can prepare themselves for and undertake passages between incommensurable systems/worlds that normally cannot come into physical contact. Most, if not all, world cultures describe such liminal situations in the form of ghost stories and other traditional accounts of encounters between the living and the dead.

The term *liminality*, as I have mentioned in chapter 2 above, comes from the Latin word *limen*, "threshold" and therefore is semantically related to *āyatana*, "passageway." Macy's "nondiscriminatory awareness," then, can more properly be described as a liminal state. It is the in-between state in which consciousness, while preserving its transitivity, perceives the ceaseless formation and transformation of objects and other self-organizing patterns without attaching itself to any of them in particular. Therefore, nondiscriminatory consciousness, like any other liminal state, does not construct alternative worlds, even though it can assist in their emergence, in which case it again becomes discriminatory. It is also beyond death, understood as the dissolution and/or transformation of self-organizing systems into other systems. By the same token, it is only discriminatory consciousness that is attached to what one conventionally understands by life and death.

The crucial question is, however, whether the Buddha also envisages the possibility of alternative worlds beyond the cycles of birth and rebirth arising within the worlds of power, or only the possibility of the cessation or the "snuffing out" of the latter, as many Buddhist and non-Buddhist scholars assume. One instance, among others, in which the Buddha appears to be fully aware of both the possibility of alternative worlds outside power and the important role of liminal interstices in preparing the conditions for the emergence of such worlds can be found in part I.IV of the *Saṃyutta Nikāya*, in the *Mara Saṃyutta*, which describes the encounters between evil Mara and Gautama the Blessed One.

In one such encounter (*Marasaṃyutta* 20.10, *Rulership*), the Buddha has retreated to the Himalayas and lives in a leaf hut, meditating on the nature of human governance. One day he asks himself if it were possible to exercise governance "righteously," that is, "without killing and without instigating others to kill, without confiscating and without instigating others to confiscate, without sorrowing and without causing sorrow." At this juncture, Mara the Evil One materializes before him out of thin air and says: "Let the Blessed One, exercise rulership righteously," because he "has developed and cultivated the four bases for spiritual power, made them a vehicle, made them a basis, stabilized them, exercised himself in them, and fully perfected them." Indeed, if the Blessed One so wishes, "he need only resolve that the Himalayas, the king of mountains, should become gold, and it would turn to gold." So why couldn't he also govern righteously? But the Blessed One does not fall for this ploy, whereupon Mara melts back into thin air.

The Buddha immediately sees through the trap that Mara has set for him. For one, Gautama is considering the issue of governance within the reference frame of an irenic world, in which systems organize themselves and correlate with other systems according to the principle of peace. The Evil One tempts him to rejoin the world of power that the Buddha has already left behind when he entered the liminal world of the Himalayas. In this power-oriented world, quite removed from both the liminal place of his leaf hut and that of the irenic world that he is imagining, another reference frame applies, where most eyes, including Mara's, are indeed "fixed on the gold," as E. O. Wilson says. Within this reference frame, "righteous rulership" would be understood in relation to power, which creates a system of values and corresponding reality that are radically different from those of an irenic reference frame.

Power as an organizing principle tends to see irenic value systems as merely incompatible with it, because it operates in terms of a polarity between war and peace. From its standpoint, therefore, peace is merely the absence or lack of war (cf. my discussion of Symonides and Singh's democratic definition of peace, at the end of part I above). It may also suit its purposes to pretend at times that war is waged for the sake of peace, as in the well-known Roman adage: "If you want peace, prepare yourself for war." Yet, the two value systems are not incompatible, but incommensurable, so that what appears as "righteousness" in one system has a completely alien meaning and is untranslatable into the other.

As we have seen, the *Tao Te Ching* equally makes this point when it emancipates the "weaker" terms from the conceptual polarities of a power system and grants them independent, amplifying feedback loops. In such loops, war generates more war, and peace generates more peace. In this connection, Lao Tzu recounts a "golden age" story about true governance according to the Tao. In the "olden days," when people were governed by the Tao, they had the best rulers, because they were not even aware that they were ruled. So that, when the ruler's "tasks were accomplished and his work done, the people said: It happened to us naturally" (I.XVII).

The next age marked a paradigm shift, incommensurable with the previous, Taoist one. Through this shift, polar thinking, acting, and being in the world, organized according to the power principle, came into existence. Referring to the consequences that this shift of paradigms had on governance, Lao Tzu notes that when "there is not enough faith, there is lack of good faith" (I.XVIII):

> Hence when the way was lost there was virtue; when virtue was lost there was benevolence; when benevolence was lost there was rectitude; when rectitude was lost there were rites.
> The rites are the wearing thin of loyalty and good faith
> And the beginning of disorder....
> (II.XVIII)

During this fallen age, therefore, people first had rulers they loved and praised, then rulers they feared, and finally rulers they took liberties with. It was also during this second age that there appeared not only "benevolence" and "rectitude," "cleverness," "filial children," and "loyal ministers" (I.XVIII), but also their polar correlatives, such as ill will, crookedness, hypocrisy, ungrateful children, and treacherous advisers.

Lao Tzu then suggests a return to the Taoist "golden age" of governance in which the polar thinking of a mentality of power will be dissolved. But from the standpoint of power, this dissolution will appear as a typical Taoist paradox:

> Exterminate the sage, discard the wise,
> And the people will benefit a hundredfold;
> Exterminate benevolence, discard rectitude,
> And the people will again be filial;
> Exterminate ingenuity, discard profit,
> And there will be no more thieves and bandits.
> ...
> Have little thought of self and as few desires as possible.
> (I.XIX)

The paradox can be removed, Lao Tzu implies, only with the ushering in of the previous, incommensurable paradigm, alluded to in the last verse of the preceding citation. Not thinking of oneself and extinguishing desires of power are also Buddhist precepts, as we can see from the Mara story in which Mara tempts the Blessed One with returning to being a ruler, even though he had renounced that role when he took up the life of an itinerant monk.

One should note that the Buddha asks himself the question about righteous governance from inside the liminal world of his leaf hut, and it is for this reason that he puts it in the negative terms of the world of power he has just left behind. He has not yet fully developed the reference frame of an irenic world, but he has now gone one step further toward it, not least because he has not succumbed to Mara's temptation. In turn, Mara sees his chance of tricking him, precisely because the Buddha is situated in a liminal world, where the reference frame of a world of power, although no longer in force, is still present, arising to consciousness along with other organizing principles and reference frames, including those of an irenic world.

5 Irenic Mentality and the Sufi Path

Finally, we can now also look upon Rumi's poem that serves as epigraph to the present book as an expression of the same nonlinear, irenic way of thinking in the Islamic tradition (*pace* Huntington).[5] The thirteenth-century Sufi poet employs the same Taoist and Buddhist strategy of dissolving the conceptual polarities generated by a power-oriented mentality, in order to reveal the irenic world of love to which he feels he belongs. He situates himself in a liminal space between these two worlds, pointing to their incommensurability. The language into which he has been born (Persian) and in which he must address his audience belongs to a power reference frame and, therefore, cannot express the concepts and values that belong to the incommensurable world of love. For that reason, he can describe this world only in negative terms, as what it is not, rather than what it is. Consequently, in the first line of the poem, he adopts the viewpoint of his Muslim audience, declaring ironically that he does not know who he is: "What shall I say, O' Muslims, I don't know myself."

In the reference frame of a world of power, everyone strives to "know" himself and others, that is, to establish his identity and place in the community, based on religion, ethnicity, birth, gender, social position, or any other differential principle. But Rumi proceeds to dissolve all such identities and differences, so that his audience can no longer pin him down and engage him in a dialectics of power. First, he claims that he belongs to no religious faith: "I am neither a Christian, nor a Jew, nor a Zoroastrian, nor a Muslim." Nor does he belong to any ethnic group, for he is neither "an Indian, nor Chinese, nor Bulghar, nor Saksin/ Neither of Iraq, nor of Khorasan." Nor does he belong to any known culture or civilization, whether terrestrial or extraterrestrial, because he is "Neither of the East, nor of the West, nor of the desert, nor of the sea/ Neither from the land, nor of the sky."

Once he has shed all cultural identities and differences, Rumi goes on to erase all of their physical and/or spiritual counterparts. He declares that he is "Neither of the earth, nor of water, nor of wind, nor of fire/ Neither of the high, nor of low, nor of space, nor of time/ Neither of this world, nor of the next, nor of paradise, nor of hell." He even dissolves his gender identity, for he is "Neither of Adam, nor of Eve." Having left behind all differential principles conceivable to a mentality of power, he situates himself in a liminal interstice, where his place is the "placeless," his sign, the "signless," and where "there is neither a body nor a soul." No polarities, identities, and differential principles remain operative within this liminal space. And now that he has led his audience literally into a no man's land, he can finally reveal his true identity, which belongs to an incommensurable reference frame, with an entirely different system of values and beliefs: "For I am of the Beloved." The world of love to which he belongs and that he identifies with God cannot be understood in the polar terms he has invoked—and dissolved—for his Islamic and other audiences. One would need to invent an entirely new language, indeed an alternative mode of thinking, feeling, and acting, in order to allow this world to emerge.

There are many other Sufi masters, before and after Rumi, who proclaim and practice what they, like their Taoist and Buddhist counterparts, call the Path. This Sufi path is no less paradoxical than the Tao or the Dharma. For example, Bayazid Bistami says: "The thing we tell of can never be found by seeking, yet only seekers find it" (Frager and Fadiman 1997, p. 37). The Sufi practitioners also advocate an equivalent of the Taoist *wu wei* or acting according to, rather than against, the nature of things. In this respect, Amr ibn Uthman al-Makki notes: "The Sufi acts according to whatever is most fitting to the moment." In turn,

Abu Sa'id urges the practitioner along the same lines: "Whatever you have in your mind—forget it; whatever you have in your hand—give it; whatever is to be your fate—face it" (ibid.). Here "fate" should not be understood as a "fatalistic" abdication of responsibility or freedom, but as action in accordance with the natural course or the Path.

As Rumi's poem implies, the Sufi masters equally employ the concept of nothingness as a strategy of going beyond a mentality of power. For example, a traditional Sufi story recounts how at a royal banquet everyone was seated according to his rank, waiting for the king to appear, when a simply dressed man came in and took a seat above everyone else. The prime minister demanded that the man identify himself, but every time the minister mentioned a rank, including that of king, the stranger declared he was above it. "Then you must be God," the prime minister concluded sarcastically. "No. I am above that," replied the stranger. "There is nothing above God!" shouted the infuriated minister. Upon which the man said unruffled: "Now you know me. That *nothing* is me" (Frager and Fadiman 1997, p. 250).

Most important for the present argument, given the current, widespread, Western misconceptions about Islam, is the Sufi attitude toward conflict, violence, and sectarianism. We have seen, in my discussion of Huntington's work, that the prophet Muhammad decries conflict and preaches meekness in the Taoist, Buddhist, and Christian sense. In turn, al-Gazzali imagines the Day of Judgment, when the Lord will summon, among many other people, those who gave away their lives in the Holy Wars: "They will be asked, 'How did you spend your life I gave you?' They will reply, 'We sacrificed it in Your Path.' The Lord and the angels will call them liars and say, 'You gave away your lives that people might call you brave and style you martyrs'" (ibid., p. 63).

According to Murat Yagan, moreover, although there are many different Sufi orders, "*there is no lack of love or respect between these various orders*. They do not reject each other, or criticize each other. Nor do they claim to be closer to the Truth. Sometimes it is said, 'The fountain from which I drank was here, and there are many other fountains if you are thirsty'" (ibid., p. 39; italics in the original). Finally, Shabistari invokes a beautiful metaphor that can also serve to convey the spirit of global intelligence, as defined in the present study:

"I" and "you" are but lattices,
In the niches of a lamp
Through which the One Light shines.
"I" and "you" are the veil
Between heaven and earth;
Lift this veil and you will see
No longer the bonds of sects and creeds.

(Ibid.)

5

Toward an Ecology of Ecology

The previous discussion has shown, I hope, that early Taoists and early Buddhists, as well as later Sufi practitioners, are fully aware that understanding mutual causality and other nonlinear processes will not by itself guarantee transcendence of a mentality of power and its corresponding worlds. As we have seen, nonlinear ways of thinking could, more readily than linear ones, facilitate such transcendence, but they are only means toward an end. Means and end, moreover, should never be treated separately, but as causally reciprocal, amplifying correlatives. In order to go beyond our currently prevailing mentalities, early Taoists and Buddhists and their later Sufi counterparts suggest that we need to develop alternative modes of being, incommensurable with power-oriented ones. In turn, such modes will generate alternative worlds, in which living systems will organize themselves and interact with other systems on principles other than power. My assumption throughout the present book has been that it is this kind of evolution that the collective efforts of humanity in all fields of activity should focus on during the next few centuries.

Should we Westerners wish to initiate the mutual causality processes that will eventually lead to the emergence of such worlds, we need to continue rethinking our current notions of self-organizing systems. We may start with the principle of hierarchical order, which remains linear in most complexity and chaos theory, despite attempts to redefine it in nonlinear terms. Laszlo, Eisler, Capra, Macy, and other systems theorists that belong to various ecological and feminist movements are fully aware that linear causal thinking is a form of hierarchical thinking, proper to a mentality of power. Macy, for example, mentions that some systems thinkers such as Arthur Koestler propose the term *holonarchy* (from the Greek *holos*, whole) to describe nonhierarchical relationships among systems and subsystems and "to avoid confusion with notions of rank or unidirectional agency" (Macy 1991, p. 77). Yet, this does not solve the problem, because it is *-archy*, the second semantic component of the term "hierarchy" (coming from the Greek *arche*, origin, first principle, determining cause, power) that qualifies the nature of the relationships among *holoi*. Such relationships necessarily remain agonistic and/or hierarchical in current general systems theory and cybernetics.

As the history of humanity has repeatedly shown, paradigmatic shifts within power-oriented systems are easier to effect than other shifts. In this respect, a transition from a linear to a nonlinear mode of thinking, or from unidirectional to mutual causality does not necessarily imply a shift of incommensurable paradigms. In most cases, it only implies a shift of incompatible paradigms, so it is not by itself "a radical shift in worldview," as Macy and other ecologists believe. It is for this reason that systems theory and cybernetics can be and have been applied to management, industry, and information technology, for purposes of efficiency, command, and control. As such, they have served to enhance a dehumanizing and self-destructive technocratic mentality. One should not forget that cybernetics was first developed for military purposes. Wiener defined it, we recall, as "the science of control and communication in the animal and the machine." He first utilized cybernetics in perfecting ballistic technology for the U.S. Army during World War II, even though later on, during the Cold War period, he became unhappy about its uses by the industrial military complex (not unlike the scientists who developed the nuclear bomb).

Nor does it suffice to say, as Macy does, that the usefulness "of systems theory has been obscured by its far more extensive application to microeconomic objectives than to macroeconomic concerns" (Macy 1991, p. 207). It is certainly true, as she notes, that in "systems analysis and management, it has been employed as a corporate tool for purposes of efficiency, with little questioning of the values and goals of this 'efficiency' within the larger biosocial systemic hierarchy" (ibid.). Yet, it could be used in a similar way at the macrolevel, for example in a totalitarian environment, where it could function very well, as Arthur Koestler and George Orwell demonstrate in their fiction.

In *The Will to Power* (1967 [1901]), moreover, Nietzsche has laid the foundation of an entirely holistic science of biology, based on the ceaseless play of physical forces that form and dissolve alliances within and among living systems, according to their insatiable appetite for power.[1] Nietzsche, Heidegger, and their postmodernist followers such as Bataille, Deleuze, Foucault, Derrida, and others can equally be seen as holistic thinkers who place nonlinear concepts in the service of a power-oriented mentality (Spariosu 1989). Finally, most archaic mentalities, such as the Hellenic one, are both power oriented and holistic (Spariosu 1991). Thus, holism in itself does not necessarily imply a radical shift in paradigms, any more than nonlinear causality does.

Another reason for the great difficulties we experience when trying to shift to an alternative ontoepistemological paradigm is that our vocabulary of power is so pervasive, as the term "holonarchy" has just illustrated, that it is hard to imagine relationships within and among systems that are based on organizing or weighting principles other than power.[2] Our physical sciences are entirely based on this vocabulary, derived from Hellenic archaic words, denoting various forms of physical force (Spariosu 1991). Terms such as "dynamics" (from the Greek *dunamis*, power), "equilibrium of forces," "states far from equilibrium," "negative" and "positive" feedback loops equally belong to this vocabulary.

Western physics itself is based on the notion of force, as physicist Max Jammer aptly points out. In his illuminating book, *Concepts of Force: A Study in the Foundations of Dynamics* (1957), Jammer draws a comparison between Western and Jaina physics in terms of their different organizing principles: "The Jainas, followers of Jina (Vardhamana), an elder contemporary of Buddha, developed a realistic and relativistic atomistic pluralism (*anekantarada*) without the slightest allusion to the concept of force, in contrast to Western science, in which the idea of force plays . . . a fundamental role" (Jammer 1957, p. 5).

Jammer necessarily fails in his attempt to translate the Jaina physical concepts into Western ones, not to mention his characterization of Jaina physics as "realistic and relativistic atomistic pluralism." As he is fully aware, there are no terms in Western physics that could properly convey Jaina words such as *pudgala* (translated as "matter"), *akasha* ("space"), *kala* ("time"), *dharma* ("motion"), *adharma* ("rest"), *parinama* ("change"), and so forth. In order to understand the full extent of the incommensurability between Western and the Jaina physical concepts, one would have to submit them to a detailed comparative analysis, in both conceptual and historical-etymological terms.

Such an intercultural comparative analysis is well beyond my purpose (and competence) in the present book, but could constitute a worthwhile research project for an intercultural and cross-disciplinary research team. My point, here, is simply that if we genuinely wish to change our current, power-oriented, scientific paradigms, we would ultimately need to develop a different scientific vocabulary, corresponding to different ontoepistemological reference frames and different forms of science, including physics and biology. An excellent beginning in biology, for example, is the concept of cooperative biological relationships that Margulis (in the wake of Merezhkovsky) terms *symbiogenesis*, although we still have a long way to go in building a consilient, general theory and practice of evolution organized around symbiotic principles.

The contemporary science of ecology, which has extensively employed concepts borrowed from systems theory and cybernetics, has made some progress toward developing a coherent theory of life sciences in which power is beginning to be questioned as a viable organizing principle of living systems. Bateson, Laszlo, Eisler, Capra, Macy, Gare, and others have described the ways in which dualistic perceptions of mind and body have created, in Macy's words, "an adversarial relationship between ego and nonego, engendering myopic pride and a kind of moral blindness we can no longer afford" (Macy 1991, p. 153).

For example, Gregory Bateson calls our mentality of power a "versus" mentality and attributes it largely to a Cartesian dualism of mind versus matter that we saw at work in Wilson's sociobiology as well. Bateson believes that if we continue to operate in terms of this Cartesian dualism, "we shall probably also continue to see the world in terms of God versus man; elite versus people; chosen race versus others; nation versus nation; and man versus environment" (Bateson 2000 [1972], p. 337). According to him, this kind of mentality will lead our species to self-annihilation. If we are to survive, therefore, we must recognize that mind extends beyond personal consciousness and "is immanent in the larger system—man *plus* environment" (ibid., p. 317). For Bateson, an ecology of mind involves going beyond Cartesian dualism, to a holistic, systemic view of mind and matter.

Macy, in turn, makes the important point that, in early Buddhism, mind becomes "liberated not by setting itself apart from phenomenality," as it happens in various Vedic and Western forms of idealism, but by "increasing its awareness of it" (Macy 1991, p. 155). In this regard, contrary to a common misunderstanding of the Buddha's teaching, the Buddhist practitioner is not asked to hate or to reject this world. For example, Macy notes that *Majjhima Nikāya* II.232 records the following dialogue between the Buddha and his disciples, in which the Blessed One asks them to imagine a man who picks at the earth with a shovel, thinking that he could make "not-earth" out of earth: "What do you think about this monks? Could that man make this earth not-earth? No, Lord. What is the reason? It is that this great earth, Lord, is deep, it is immeasurable, it is not easy to make it not-earth before that man be worn out and defeated."

According to I. B. Horner, if the last words of the foregoing citation were translated literally, they would read: "that man would be a partaker of exhaustion and slaying." This would suggest, as Macy writes, that "violence is, reciprocally, bred in us by that which we inflict on nature and body. We assault ourselves" (Macy 1991, p. 154). Therefore, the Buddhist practitioner, no less than the Taoist or the general systems thinker, is not asked to reject or to abandon the "material" world, but, on the contrary, to engage with it in constructive, rather than destructive, amplifying feedback loops. Any retreat from "the world," as we have seen in the Buddha's encounter with Mara, is a temporary, liminal one, allowing the practitioner to stand back from and reflect on the present state of affairs, so that she can devise appropriate ways of transforming herself and the world that co-arises with her.

Yet, the ecological movement has a long way to go, not least because it continues largely to operate with a dualist notion of power, which is expressed, in political terms, through a binary opposition between "domination," or power "from top down," and "empowerment," or power "from down up." The first is usually identified with linear, hierarchical forms of power, whereas the second is associated with supposedly nonlinear forms, such as resistance or liberation movements, grassroots or citizens' movements, and so forth. Needless to say, domination is considered ethically repulsive, whereas empowerment is regarded as ethically desirable.

Macy, for example, writes that her work has been shaped "by certain moral and philosophical biases concerning the source of values and the locus of power." According to her, a hierarchical, unidirectional view of reality and causality attributes value and power to "an absolute entity or essence," assuming that "power works from the top down" (Macy 1991, p. xiii). While studying Buddhism and general systems theory, she claims to have "uncovered a radically different perspective on the source and nature of power," where "order is not imposed from above," but rather is "intrinsic to the self-organizing nature of the phenomenal world itself" (ibid.). This kind of power, which she calls "empowerment," works from the grass roots up and attempts to deal with the "powerlessness" that most people feel in the face of power that comes from above (ibid., pp. xiv–xvii).

Macy also associates empowerment with "deep ecology," a concept developed by the Norwegian philosopher Arne Naess, one of her friends and coworkers in the ecology movement.[3] "Deep ecology work," as the practice of deep ecology has come to be known since the publication of *Thinking Like a Mountain: Toward a Council of All Beings* (1988), a book Macy coauthored with Naess and other fellow ecologists, "seeks to expand the notion of self beyond the confines of ego and personal history, and to extend concepts of self-interest to include the welfare of all beings" (Macy 1991, p. xvii). Although I obviously share some of the goals of deep ecology, I am not sure that "empowerment" and "self-interest," even in an extended or enlightened form (cf. my discussion of Peter Singer in the introduction), would be the most appropriate ways of reaching such goals.

One should not forget that, according to the Buddhist and Taoist insight regarding amplifying feedback loops, power as an organizing principle will invariably engage in a destructive cycle, whether it comes from above or from below. As I have repeatedly argued throughout this book, power feeds and thrives on empowerment and resistance without which it simply cannot operate. It is precisely the absence of empowerment and resistance that it fears, qualifying it as "nothingness" or "emptiness." And it is for this reason, as we have seen, that Taoist, Buddhist, and Sufi thinkers cultivate emptiness as a strategy of turning away from power and toward worlds organized on other principles.

The ecological notion of empowerment is grounded in the belief that the perspective of the exploited or the victim is somehow more redeeming than that of the exploiter or the victimizer—a belief that is present in many other Western ideologies, ranging from Marxist to Christian dogma. Attempts to build equitable communist and other societies based on this belief have, however, proven disastrous, largely because the perspective of the victim and that of the victimizer go hand in hand, according to the mutual causality processes described by early Buddhism and Taoism. Among contemporary deep ecologists, Wendell Berry and Paul Hawken are equally aware of this amplifying feedback loop of power. For example, Hawken cites Berry's observation that no matter how destructive governmental policies and corporate methods and products may be, "the root of the problem is always to be found in private life." Every issue that concerns us, Berry adds, "leads straight to the question of how we live. The world is being destroyed—no doubt about it—by the greed of the rich and powerful. It is also being destroyed by popular demand. There are not enough rich and powerful people to consume the whole world; for that, the rich and powerful need the help of countless ordinary people" (Hawken 1993, pp. 14–15).

Nor is it enough to profess, like Wilson, to be a "holistic" scientist, or a "deep" ecologist concerned with the environment. Unfortunately, his case is far from being an isolated one. Even a quick glance at the history of the ecological movement in the past fifty years or so shows that this movement is deeply problematical.[4] C. S. Holling (1995, pp. 3–34), among others, has examined the conflict between the "two streams of science," which I have also staged in chapter 3 above. According to him, one stream is experimental, reductionist, and disciplinary, representing what I have called the "Western-style, mainstream scientific paradigm." Holling offers molecular biology and genetic engineering as prime examples of this kind of science. The other stream, less familiar to the general public, is interdisciplinary, holistic, comparative, and historical. It employs multiple-scale analysis and is also experimental, but at appropriate scales: it does not require disproof by experiment or the official seal of approval of the mainstream scientific community. Holling cites evolutionary biology and systems approaches in populations, ecosystems, landscapes, and global dynamics as examples of this second stream of science (ibid., pp. 12–16). Predictably, he is more sympathetic to the second scientific approach than to the first one, just as I am.

When translated into ecological terms, the two streams of science support and justify, on the one hand, "shallow" (Arne Naess's term) or mainstream ecology, of which Wilson is one of the leading theorists and promoters, not only in *Consilience*, but also in his most recent book on *The Future of Life* (2003); and, on the other hand, deep ecology, whose advocates include, as we have seen, Gregory Bateson, Ervin Laszlo, Paul Hawken, Wendell Berry, Fritjof Capra, Joanna Macy, and Arne Naess, among others.

In the political arena, the relations between practitioners in the two camps are often strained, with mutual recriminations and little desire for genuine dialogue. Both camps are concerned about the environment, but are bitterly divided regarding environmental objectives and the means of attaining them. The conflict between them tends to create confusion among politicians and the general public, being as acrimonious as any other civic disagreement within a power-oriented reference frame. Both camps have their fundamentalists and extremists who believe that any means, including disinformation campaigns, political blackmail, protracted legalistic wrangling, and violence will justify a noble end. They often engage in destructive, amplifying feedback loops with their ecologically incorrect opponents, thus contributing to a further deterioration of the global human and natural environment. They have also occasionally switched sides, "selling out" to multinational corporate interests.

What the two streams of ecology, therefore, have in common is an agonistic concept of nature (including humanity) as a ceaseless play of physical forces, out of which "ecosystems" emerge as a precarious and ever changing balance among such forces. It is at this point that the ecology based on systems theory and one based on early Buddhist, Taoist, and Sufi thinking part ways. It is true that these ancient schools describe earthly existence, as we presently know it, in dynamic, agonistic terms. Yet, they also wish to turn away from the present state of affairs, suggesting alternative ways of organizing all life on this planet. General systems theory, on the other hand, largely contents itself with conserving the status quo and shies away from a long-term evolutionary view. In the long run—speaking in geological time—evolution might well take the direction envisioned by the early Taoists and Buddhists, especially if humanity collectively makes the choice of moving in that direction.

Even those deep ecologists who, like Bateson, Laszlo, Capra, Macy, are fully aware that environmental issues cannot be separated from an overall ecology of human relations, remain within an agonistic mentality, not only in matters related to political activism, but also in their scientific beliefs. For example, they continue to place undue emphasis on such evolutionary notions as adaptability, survivability, and competition, albeit they would now approach them in nonlinear rather than in linear terms. A good illustration of this misplaced emphasis is Laszlo's concept of "adaptability" in social systems.

Laszlo suggests that if a social system becomes dysfunctional, violent corrective measures might be necessary. According to Laszlo, a social system becomes dysfunctional when it hinders diversification and the free flow of information through the coercion of its members, or when it cannot integrate its members into the larger systemic hierarchy so as to live in harmony with other social or natural systems within that hierarchy. For him, to "adapt" to such social systems is no more desirable than "to 'adapt' to a tumor on the brain" (Laszlo 1973, p. 273). Whether the corrective measures involve "reform or revolution" depends on the flexibility of the particular system. If it is too inflexible to respond to progressive therapy, then radical measures may be needed to "readapt" it to the larger systemic hierarchy (ibid., p. 274). Yet, this kind of "shock therapy," whether one deals with a malfunctioning social system, a malignant tumor on the brain, or a terrorist organization, has very seldom worked in the past, and it is part of the destructive amplifying feedback loop, in which violence always breeds more violence, as Lao Tzu pointed out more than two and a half millennia ago.

Laszlo's scientific approach to social organization is particularly dangerous when he applies general systems theory to what, ironically, he calls "world order." For example, in *A Strategy for the Future: The Systems Approach to World Order* (1974), Laszlo proposes a "world homeostatic system" as a mechanism of solving the most recalcitrant and urgent problems that plague our planet. His call for the development of global consciousness and for the integration of national and regional perspectives into a global one, as well as his analysis of the social groups and organizations most likely to work toward a global society are certainly valuable and remain useful today. Yet, his global model of a "central guidance system," based as it is on advanced Western technology and bureaucratic expertise, could easily become a dystopian nightmare. Indeed, many of its features have already been put in place since Laszlo published his book a quarter of a century ago, without bringing us any closer to the desired goal of transforming a growth-oriented, materialistic society into an equitable one.

There are at least four serious problems with Laszlo's model. The first one is that it addresses solely the physiological and safety needs of the world communities, for the basic, utilitarian purpose of ensuring the "survival of the species." Laszlo's central guidance system would thus regulate human behavior in the areas of security, economy, ecology, and population growth. Less "basic" needs, such as emotional, cognitive, and aesthetic ones, can be entrusted to national and local governments, or be left to their own devices. The second problem, closely related to the first one, is that the central guidance system relies heavily on Western advanced technology, defined as "all forms of purposive human control over the natural environment" and as the engine of sociocultural development. Laszlo justifies this heavy dependence on technology by appealing to his notion of feedback loops. Societal organization tends, in his view, toward stability and conservatism and therefore involves negative feedback loops, or Cybernetics I. By contrast, new technologies destabilize social structures, moving them toward new levels of organization through positive feedback cycles or Cybernetics II (Laszlo 1974, pp. 37–38).

By privileging human material needs over immaterial ones, as well as by separating the human environment from the natural one, Laszlo unwittingly falls into the dualistic and utilitarian views of reality that he critiques in the case of reductionism. Furthermore, by viewing technology as a form of human control over the natural environment, he not only perpetuates the split between nature and culture, but also neglects the ideological nature of his technoscientism. He thus believes that his technoscientific central guidance system can be ideologically neutral and exercise "a measure of control over critical global processes without having to deny different political-ideological systems the right of existence" (ibid., p. 59). According to him, a "socialist technology pollutes neither more nor less than a capitalist one, nor does it differ in its potential to harvest food and natural resources." Therefore one would not need to enforce "common organizing principles on politically constituted entities—it is sufficient to prevent them from stressing the carrying capacity of the global environment beyond levels where human qualities of life suffer irreversible damage" (ibid.).

The third serious problem, equally linked to the others by mutual causality, concerns the organization of the central guidance system. On the one hand, according to Laszlo, the appropriate global decisions would be reached not from top down, but from the grass roots up, through town hall meetings or their equivalents at the various systemic levels (as if this were possible in autocratic societies, whether communist or not). On the other hand, the same global decisions would be implemented, from top down, by a hierarchy of global technocrats. Even assuming that such technocrats would be selected because of their high-minded, globalist values and their commitment to environmental and economic fair play, the question remains how they would be able to carry out their mandates in a politically and ideologically divided global arena.

The fourth problem, amplified through reciprocal feedback from the other three, is the enforcement mechanism that the central guidance system would be entrusted with. The supranational bureaucratic elite would have at its disposal a nuclear force impressive enough to deter any single powerful or "rogue" state, or paramilitary global organization, from infringing on the planetary code of ethics established by democratic process at various systemic levels. It is clear, even from this brief, admittedly schematic account, that Laszlo is far from renouncing or even seriously questioning a mentality of power that permeates the very fabric of general systems thinking.

For his part, Arrin Gare, who has offered the most far-ranging critique of the mainstream Western mentality and its "culture of nihilism" to date and who, like me, does not exempt the deep ecologists from this critique (Gare 1993a, pp. 55–60), is no more prepared to give up the notion of power than any of his post-Marxist colleagues is. For example, in the introduction to his fine book, *Nihilism Incorporated*, he notes that the acute global environmental crisis in our time is useful in revealing "not only the total inadequacy of prevailing social thought, but also the nihilistic attitudes which dominate the modern world." This crisis equally reveals "the imperviousness of governments to the obvious irrationalities of the present economic order, and their unwillingness to do anything to seriously tackle the world's long-term problems; and underlying this, a political world dominated by politicians, press barons, corporation chiefs, bureaucrats, military and intelligence moguls, devoid of ideals and principles, for whom the struggle for power has become a mere sport" (ibid., p. 2). Here Gare implies that there is also a struggle for power that is more than mere sport; say, "serious," post-Marxist activism.

Indeed, Gare believes, just like the rights theorists, utilitarians, mechanical materialists, reductionists, and other latter-day Hobbesians whom he criticizes, that power *is* a natural, first principle. He clearly expresses this belief, for instance, in his post-Marxist concept of "praxis": "By focusing on praxis, the issue becomes what are people's ultimate ends in life. In according with both the metaphysics I am defending [process philosophy] and with the achievements of the human sciences, I am proposing a conception of humans which implies that they are striving to gain power, to gain recognition of their significance, and to orient themselves in the world" (ibid., p. 63).

According to Gare, these are absolute, anthropological universals. In turn, what is "culturally relative is how people strive for these ends. To understand the role of culture it is necessary to see it in relation to socially situated praxis directed towards these ends" (ibid., pp. 63–64). Gare seems unaware of the fact that two of the human ultimate ends he describes, namely, recognition of personal significance and orienting oneself in the world are directly subordinated to the first one, that is, striving to gain power. Consequently, Gare also seems unaware that power is indeed a form of sport or play, as Nietzsche and Orwell, among others, have conclusively demonstrated. As such, power's objective is nothing but an enhancement of the feeling of power, and ideals and principles are no more than instruments in achieving this objective.

Gare wishes to replace the Western nihilist tradition with another "grand narrative," based on a "consistent reformulation of Marxism in terms of process philosophy" (ibid., pp. 3–4). His recipe for radical social change is a warmed over, post-Marxist notion of Gramscian hegemony, environmentalist nationalism, and (Western-style) creative rationality, all of them grounded in process philosophy and general systems theory (Gare 1993b, pp. 214–235). But Gare might wish to consider that all of these notions, no matter how revamped, are nothing but instruments of a mentality of power that will reproduce itself by any means, including post-Marxist, environmental programs.

Unless we reorient our striving to ultimate ends other than power and thus separate our need for recognition and for orienting ourselves in the world from the power principle, we will continue to reproduce and amplify the Western nihilistic tradition that Gare analyzes so well. As I have emphasized throughout the present study, what we need at this juncture in our Western (and human) history is certainly not another "grand narrative." Rather, we need to design and implement "small" narratives or local–global cultural blueprints, grounded in a mentality of peace and an emergent ethics of global intelligence.

In turn, Macy automatically preserves the general systems' vocabulary of "dynamic equilibrium," "opposing forces," "survival," and "adaptability," even though her call for a nonviolent ecology in consonance with early Buddhist teachings renders this vocabulary inadequate. For example, she writes that by its "transactions with the surrounding world the system maintains its dynamic equilibrium in the interplay of opposing forces" (Macy 1991, p. 111). According to her, as well as all other general systems theorists, the process of evolution is "deviation-amplifying in a number of ways, featuring positive feedback loops between mutations and environment, within species and between them" (ibid., p. 98). To make her point, she invokes the well-known example, in evolutionary literature, of the ways in which prey and predators interact: "The moths with more effective camouflage survive better, as do their predators with sharper detectors. In the course of generations the moths' camouflage becomes ever more cryptic, while the acuity of the hunters' senses likewise increases" (ibid.).

In support of this thesis, Macy further cites Stephanie Hoppe who, in a collection of fiction she has co-edited about perceptions of animals, writes that prey and predator "are both forming each other. To be a predator one needs to be obsessed, one is totally formed in the direction of one's predation; on the other hand, to be prey is such an interesting existence" (ibid., p. 103). From this same book, Macy invokes a "fanciful" story by Judy Grahn as another example of the positive, amplifying feedback loop between predator and prey: "Because of her habits of trembling without moving the mouse has taught the owl to do a great deal of sitting and dreaming it would not otherwise be doing. Because the mama mouse has habits of working at night, the owl has been given plenty of night vision and silent, rapidly dropping flight in order to fall down on her from above.... For a mouse to have so many babies at a time as she loves to do she must have someone to give them over to and for this reason her kind have thought up the family of owls and dreamed them into existence" (ibid., p. 104).[5]

But these are precisely the kind of fanciful stories that our children can do without. They remind one of philosopher Pangloss's optimistic reasoning, in Voltaire's *Candide*, according to which, in the best of all possible worlds, any unfortunate state of affairs eventually works out for the best. My point, obviously, is not that prey and predators do not exist in our world, or that they do not engage in a "positive," amplifying feedback loop, especially at feeding time. But, they are certainly not "made for each other," as Grahn appears to suggest. This is tantamount to saying that the bay of Lisbon is made for drowning (as Pangloss reasons), the frog is made for the snake, the pig, cow, and sheep are made for human consumption, the bullet or bomb is made for its human target, and so forth. It takes a perverse imagination to say that "mama mouse" dreams up and confides her offspring to the owl (and many other predators) as a form of mouse population control. *Mutatis mutandis*, this idea sounds like Jonathan Swift's "modest proposal" (which he did not make in earnest, however), that child-burdened, indigent parents in Ireland sell their offspring to be served as delicacies at the table of the rich, so as to solve the problem of overpopulation and dire poverty in the British Isles.

Yet, unfortunately, it is accepted fact among ecologists, whom Hoppe and Grahn mimic, that the dynamic balance of opposing forces within an entire ecosystem depends on the interactions between predators and prey. This might be the case, if we study such interactions in isolated linear, or even causally reciprocal, binary models. But we do not know enough about the overall, long-term biological effects (in geological time) of self-organizing systems that specialize in preying upon others to draw any definitive scientific conclusions about this and many other issues that relate to the "struggle for existence," "rate of reproduction," and the "survival of the fittest." (We do nevertheless have the telling example of one species, *Homo sapiens*, whose predatory habits have gone completely out of control, to the point that it has endangered biodiversity and the quality of life on the entire planet.) In the absence of such knowledge, it might be much more productive, for the future development of life on earth, to encourage and support evolutionary hypotheses and research, such as those of James Lovelock and Lynn Margulis, which look at evolution as a nonviolent, symbiotic and creative process.

From the standpoint of an emergent ethics of global intelligence, there is nothing to lose and everything to gain in assuming that future long-term evolutionary trends could move further and further away from the age of dinosaurs and could eventually phase out violent ecosystems altogether. This phasing out would not exempt *Homo sapiens* that, in its present state of development, remains the most vicious and destructive species of predators that has ever walked the face of the earth, including the much maligned dinosaurs. But, of course, there is no reason why we humans should not work collectively toward transvaluating *Homo sapiens*, which is still at a very early evolutionary stage, so it can eventually fulfill the vocation of its name through an evolutionary bifurcation or paradigm shift of the kind that I have imagined here. We should therefore continue dreaming up, not more owls, as "mama mouse" supposedly does, or more central global guidance systems, but an alternative, irenic evolution, and then begin to work patiently, from generation to generation, toward making it a reality.

I should emphasize that I hardly mean to belittle the efforts of ecologists like Bateson, Naess, Macy, Laszlo, Capra, Gare, and others who attempt to transcend our agonistic and disciplinary mentalities, despite the fact that these efforts may often fail (and I certainly do not exempt myself from such failures). But the self-organizing systems described by them—unlike those envisioned by the early Taoism and Buddhism and later Islamic Sufism—do belong to power-oriented worlds and should be recognized as such, if we are to move forward. None of our efforts will go to waste, however, because it is crucial to understand fully the organizing patterns, structures, and processes of power-oriented systems. In fact, some of the best descriptions of such systems come precisely from those thinkers who wish to find alternatives to them, such as Lao Tzu, the Buddha, Pythagoras, Plato, the Christ, Francis of Assisi, Rumi, and numerous other sages throughout world history.

Judy Grahn's story mentioned above may constitute a good opportunity to revisit the notions of feedback loops, as systems theorists and cyberneticists currently conceptualize them, particularly if we allow Taoist, Buddhist, and Sufi thinking to resonate with us. Perhaps, we should eventually replace the technical terms of "positive" and "negative" feedback loops (which, in any case, are confusing and counterintuitive) with "productive" or "life-enhancing" and "destructive" feedback loops, which can in turn be seen as either self-regulating or amplifying. This way, we would, like the Taoists, decouple the asymmetrical polar structure of many of these feedback circuits, including those of predator and prey. Consequently, only the dominant pole will remain within the reference frame of power, with its ceaseless interplay of opposing forces, whereas the "weaker" pole will transform into the organizing principle of a different mentality altogether. In this alternative frame of reference, the notion of equilibrium (of opposing forces), no less than that of a "state far from equilibrium," would be rendered inoperative, so that different, nonviolent concepts of and relationships among self-organizing systems and other phenomena could emerge over time, at all levels.

The notion of resonance, which I have explored in chapter 2 above as a useful conceptual tool in developing intercultural studies, might help us out in the present context as well. First of all, we can now define resonance as an amplifying type of feedback loop, rather than a self-regulating type. This would allow us to understand how self-organizing systems interact with other such systems, whether they are nested within or host those systems; how certain "local" characteristics can spread so rapidly throughout systemic wholes; and how apparently insignificant events may trigger catastrophes (in René Thom's sense) within a given system or polysystem.

We can also contrast the notion of resonance with that of mimesis in relation to constructive and destructive types of amplifying feedback loops. Mimesis, translated as "mimicry" or "imitation," often involves an asymmetrical relation of power that can lead to amplifying destructive loops. It can therefore be seen as mostly a destructive form of amplifying resonance that feeds on conflict. Therefore, mimesis can be regarded as a form of conflictive or violent resonance. In turn, there are cooperative or symbiotic forms of resonance that generally lead to productive types of amplifying loops, at least in reference frames that are not power oriented.

By way of example, we may take a second look at the deep ecologists' idea, already mentioned in the introduction, according to which human ecology should "mimic" or imitate natural cycles. Hawken, for instance, calls for a "prosperous commercial culture that is so intelligently designed and constructed that it *mimics nature at every step*, a symbiosis of company, customer, and ecology" (Hawken 1993, p. 15; my emphasis).

This notion of "biomimicry" (Benyus 1997) is fast becoming an ecological mantra in the current global environmentalist movement. It is true that by the commercial "mimicking" of nature Hawken understands mostly imitating nature's way of recycling waste. But, *resonating with nature* might be a more appropriate concept to convey the reciprocal causality between humans and other living systems that should underlie the complex ecological feedback cycles or the web of life that is our planet.

Mimicry in the physical environment involves, as a rule, an element of deceit, either to attract prey or to ward off an enemy, or to use the services of a guest or a host for one's own ends, without necessarily responding in kind. The antonym of mimicry in this sense would be symbiosis or cooperation for mutual life enhancement or benefit. So, Hawken may not wish to confuse the two, as he does in the preceding citation. In fact, there are plenty of commercial organizations that mimic ecological practices for their own ends. There are also plenty of scientists and nonscientists who, like Wilson, are exemptionalist wolves in environmental sheep's clothing, to invoke again the most familiar example of mimicry from our folktales.

The concept of mimicking nature, moreover, belongs to the linear, dualistic vocabulary that regards nature and humans as independent of each other. Those environmentalists who view nature as pristine and innocent are no less dualistic than those who see it, in the manner of Wilson and other survivalists, as a theater of war, "red in tooth and claw." There are no pure, unadulterated natural processes any more than there are objective human observers. So, strictly speaking, ecological preservationism or conservationism is an illusion (even though it is probably a helpful one, at least for the time being, given the current excesses on the side of irresponsible use of natural resources). In mimicking nature we simply mimic ourselves. As Macy aptly notes, "our imaginations erect Pentagons and Disneylands, and even the land itself mirrors back our fantasies, as, gouged and paved over, it testifies to our search for mastery and our fear of what we cannot control. In the world we create we encounter ourselves" (Macy 1991, p. 126).

Put in a different way, if we mimic nature, nature will mimic us. Depending on how we approach it, it will reflect back to us either our violent, predatory behaviors and wasteful practices, or our gentle, restorative and life-enhancing ones; or, more often than not, both. If, as early Buddhist, Taoist, and systems thinkers believe, nature and culture are interdependent, engaged in causally reciprocal processes, then these processes will resonate with and amplify each other for better or for worse. Nature can learn from us as much as we can learn from nature, so that we should look for the most creative courses of development within both the cultural and the natural sphere, to the benefit of both.

In developing an appropriate vocabulary for an alternative value system, one should also phase out the notion of survival of the fittest, or any other, living systems. This notion plays a central role within feedback loops that are based on the organizing principle of power. As it stands now, all ecologists, whether deep or mainstream, use the concept of survival both to describe nonlinear feedback cycles and as an existential imperative to deter humanity from the ecologically destructive course that it is currently embarked on. This use is predicated on the utilitarian assumption that life constitutes a supreme value in itself and that staying alive is the overriding goal of all living systems.[6]

We have seen that Laszlo bases his entire world homeostatic system, backed by nuclear deterrence, on the notion of survival. In his wake, Macy writes with regard to globalization: "Social forms of consciousness must emerge *if this planetary experiment is to continue*. Our political and economic interdependence may have progressed to a degree where collective self-awareness must manifest itself *for the world as we know it to survive*" (Macy 1991, p. 202; my emphasis). She further echoes Laszlo in saying that because a world system "in which we all inhere" is emerging, our "*adaptation and survival* require higher allegiances than those we accord nation-states" (ibid., p. 202; my emphasis). In turn, Capra, after elaborating on David Orr's concept of "ecoliteracy" (Orr 1993), concludes: "These, then, are some of the basic principles of ecology—interdependence, recycling, partnership, flexibility, diversity, and, as a consequence of all those, sustainability. As our century comes to a close and we go towards the beginning of a new millennium, *the survival of humanity will depend on our ecological literacy*, on our ability to understand these principles of ecology and live accordingly" (Capra 1997, p. 295; my emphasis).

I certainly agree with Laszlo and Macy that we need to develop collective self-awareness and go beyond national allegiances in the present global circumstance. I equally agree with Orr and Capra on the principles of ecological literacy, even though I am not very happy with this term, any more than I am happy, for the reasons I have already discussed, with the term "partnership." Ecoliteracy might be a catchy political slogan in the West, but it leaves out a large portion of the world's population that may be "illiterate" and yet, environmentally speaking, is much more "advanced" than many Westerners who supposedly know how to read and write. Be it as it may, I do believe, however, that the reforms envisaged by Laszlo, Macy, and Capra should not be predicated on adaptation and survival, but on a value system that goes well beyond such concepts.

It is obvious, for instance, that if humanity decides to develop along the lines of a mutual morality of nonviolence, which Macy and Capra support just as I do, then the world as we know it will certainly not survive, but will be replaced by an alternative world, incommensurable with the first one. It is for this reason that, at least for the time being, it would be wiser to steer world communities, gradually and patiently, toward such irenic principles and practices, primarily through education, intercultural dialogue, and cooperative learning projects, than to indulge in any scientific or political schemes for a New World Order.

If we wish to revaluate life, as well as death, in consonance with an organizing principle other than power, we should not automatically adopt the definitions of these terms that are operative within power-oriented reference frames. In such reference frames, life and death are binary oppositions, locked in a destructive feedback loop. Moreover, it is not life, but death that assumes the dominant role in these binary correlations, despite power's claims to the contrary. Hence the veritable obsession with death in contemporary Western and other cultures, that is to say, the death of others, coupled with pathological fear of one's own. This pathological fear that power shares, cultivates, and manipulates in human societies is an integral part of its either/or logic, according to which "you shall (or shall not) do this, or else you shall face the consequences," including immediate or eternal punishment, torment, and annihilation.[7]

Therefore, it is inappropriate and, more to the point, ineffective for ecologists, ranging from Wilson to Laszlo to Hawken to Capra to Macy to Gare, to invoke survivalist fears in attempting to persuade the powers-that-be and everyone else to reform our current behavior. Threats of collective extinction based on ecological or pacifist arguments do not seem to be working any better than their religious and political counterparts did in earlier times. This is the case partly because of the exemptionalist attitude of the individual human being in a power-oriented reference frame, for whom these threats and warnings apply to everyone else but himself, and partly because power itself goes beyond a survivalist mentality, even as it uses it for its own purposes. One may in this respect recall Achilles' choice, in Homer's *Iliad*, of a short and violent, but heroic life over a long and peaceful, but mediocre existence.

The issue, then, is not one of survival, as our well-meaning environmentalists and pacifists never tire of telling us, but one of transvaluating our value systems to the point where such questions become irrelevant. To begin with, we need to uncouple the two poles of the binary opposition of life and death (no less than that of war and peace) and value each process according to its specific function in the ceaseless flow of phenomena. We have seen that in early Buddhist thought, life is no longer controlled by death, nor the other way around. The same insight is present in the *Tao Te Ching*, where Lao Tzu expresses it in his usual paradoxical form, reversing and then uncoupling the polarities of life and death:

> When going one way means life and going the other means death, three in ten will be comrades of life, three in ten will be comrades of death, and there are those who value life and as a result move into the realm of death, and these also number three in ten. Why is this so? Because they set too much store by life. I have heard it said that one who excels in safeguarding his own life does not meet with rhinoceros or tiger when traveling on land nor is he touched by weapons when charging into an army. There is nowhere for the rhinoceros to pitch its horn; there is nowhere for the tiger to place its claws; there is nowhere for the weapon to lodge its blade. Why is this so? Because for him there is no realm of death. (*Tao Te Ching*, II.L)

Life can, then, be seen as a series of transformations of self-organizing systems that are in continuous flow, whereas death can be seen as a process of transition that allows the next radical reorganization of a system to take place. Both of them are important evolutionary processes, in which the notion of survival, coupled with the notion of adaptability, has its proper function as an element of temporary stability and continuity, but is not the be-all and end-all of evolution. Evolutionary theory and research practices should therefore put survival and adaptability in their proper place within the evolutionary scale and pay more attention to other, hitherto relatively neglected evolutionary factors, such as creativity, love, generosity, altruism, curiosity, play, symbiosis, and so on.

Although, as I have repeatedly said, we seem still to be far away from a radical shift in human mentalities, we can at least specify, as I have done here, what this radical shift would (or would not) involve, as well as some of the steps we can take in order to facilitate the emergence of an alternative paradigm. But, as I have pointed out in the introduction, the deciding factor in human development is not so much how effective a scientific theory might be, or how close it might be to the truth, or how easily it can be turned into a "powerful" technology. Rather, it is our daily praxis or ethos (even as we might wish, in this regard, to stay away from Gare's post-Marxist notion of "praxis" as striving for power). Traditional wisdom, which is primarily concerned with developing a proper way of life rather than an abstract philosophical system, is fully aware of the gap between theory and practice, which is just another form of exemptionalism, old or new.

In this connection, the Sufi master Ansari, for example, says: "Look to what you do,/ for that is what you are worth./ True labor means neither fasting nor prayer" (Frager and Fadiman 1997, p. 46). Another Sufi master, al-Suhrawadi, notes: "If words come out of the heart, they will enter the heart, but if they come from the tongue, they will not pass beyond the ears" (ibid., p. 39). Echoing the Buddhist and the Taoist emphasis on ethical practice rather than theory, al-Suhrawadi goes on to say: "Whoever hears something of the Sufi doctrine *and practices it*, it becomes wisdom in their hearts, and people who listen to them will derive benefit from it. Whoever hears and does not practice, Sufism is mere talk which they will forget after a few days" (ibid.; italics in the original).

Thus, one might possess the "best" scientific theory, such as deep ecology, the "best" information technology, such as digital and quantum computers, or the "best" political system, such as democracy, and yet put them all to inappropriate and counterproductive uses, as we have seen happen, again and again, in both the history of scientific and technological development and the history of world politics. By the same token, one might be a reductionist in science and a royalist in politics, and yet behave in a manner that is beneficial to everyone, rather than exclusively to one's own camp. In this regard, our position should be the opposite of Steven Weinberg's: rather than siding with the fundamentalists from any ideological camp, because they believe in their truths as strongly as we believe in ours, we should side with those from any camp who might profess less faith, but are open to genuine dialogue and are willing to work together toward the common good of all.

One should also keep in mind that in human development the means are no less important than the end and cannot be separated from the latter. If, for example, you value life and disagree with abortion practices, or if you distrust modern technology or wish to see it put to better uses, there are much more appropriate and effective ways of reaching your goal than by bombing an abortion clinic or a nuclear facility, or by sending booby-trapped mail to your colleagues in physics and computer science. Macy rightly emphasizes the "moral anguish that arises when worthy objectives seem only attainable by acts which appear, by their nature, to compromise them" (Macy 1991, p. 208). She questions the instrumentalist dichotomy between the pragmatic and the moral, in which means are subordinate to ends.

One may add that, as is often the case with power-driven polarities, "means" actually constitutes the dominating pole in this binary opposition, despite power's claim to the contrary. For example, a communist or another type of authoritarian regime, or even the neoliberal government of a Western-style democracy, may claim that it employs repressive, violent means only temporarily, in order to carry out its noble goal of an equitable or a "free" society. Yet, this goal never seems to come any nearer, and the equitable or free society always remains just around the corner.

By contrast, for Macy and for other systems theorists, as well as for the Buddhist, Taoist, and Sufi practitioners, means are "ends-in-the-making" (ibid., p. 209). In this view, ends are not preexisting, but emergent. They both shape behavior and are, at the same time, shaped by it. In other words, ends and means are engaged in causal reciprocity, generating constructive or destructive, amplifying feedback loops. As we have seen, it is in this sense that we should understand global intelligence as well: it is not something pre-given, but an end-in-the-making, informing all of our life choices at the same time that it is continuously modified by such choices.

If we accept this insight, then it is clear that scientific endeavor should equally be guided by an aspiration toward global intelligence, rather than solely by theoretical or pragmatic considerations. As Macy puts it, "the views we hold are not distant from us in time or space, but present realities, unfolding out of the core of our existence and capable of transforming it in the present moment" (ibid., p. 210). No statement, including the scientific kind "is exempt from the particularity of experience. As co-creators of our worlds we cannot extricate ourselves to claim a vision of its workings that is aloof from our own participation in it" (ibid., p. 196).

Perhaps one of the major current stumbling blocks in the development of science, as well as in human development as a whole, is our exemptionalism, coupled with self-righteousness. We have now elaborated a wide array of sciences, including an advanced, interdisciplinary science of ecology, whether deep or mainstream. But the most difficult goal is still ahead of us, namely the development of an ecology of science, including an ecology of ecology as an integral part of it. Without such ecology, we shall certainly not be able to advance beyond our present mentalities, no matter how "powerful" or sophisticated our scientific methodologies and technological gadgets will become.

It is in the name of this ecology of science that I have undertaken a critique of Wilson's sociobiology in chapter 3 above. His scientific ideas can certainly be modified and refined, but what is more difficult and far more important is to get him, and other outstanding scientists that our Western communities have produced and trained, to become aware of and reform their hubristic, exemptionalist mentality. If Wilson and, for that matter, Weinberg were to undergo this kind of conversion, then they would use the influential platform of their distinguished position in the American scientific and academic community to work, together with other scientists, on drawing up a detailed program of precisely how one can evolve from a science of ecology to an ecology of science, in which utilitarian and pragmatic questions of success, status, patronage, and power take second place to concerns for the well-being of the entire planet.

It is not enough to sound the ecological emergency alarm, as Wilson does in the last chapter of *Consilience*, and then continue business as usual. Paul Hawken and many other successful businessmen and businesswomen have had the courage and vision to propose and implement creative alternatives to the ways in which we presently conduct our commercial enterprises. It is high time for prominent senior scientists such as Wilson and Weinberg who need no longer fear the severe sanctions of the mainstream scientific establishment, to come out and teach us (as well as themselves) how to conduct our scientific enterprises in such a way that both Western science and Western societies—indeed, all human societies—can prosper, without eventually harming ourselves, our children, and the rest of the world. Wilson at least makes an honest effort in his latest book, *The Future of Life* (2003), which is an impassioned plea for ecologically responsible behavior, not least on the part of ecologists. But, again, the concrete solutions and programs he offers remain largely cosmetic or "mainstream" and do not involve a genuine shift in our prevailing mentality or in the way in which we currently practice science.

An ecology of science would involve a globally intelligent system of values, including those advocated in early Taoism and Buddhism or, later on, in Islamic Sufism. They need not be expressed only in negative terms, such as absence of exemptionalism, hubristic arrogance, and violence, but also in positive terms, such as responsive understanding and action, attentive awareness, generosity, peacefulness, benevolence, gentleness, kindness, modesty, playfulness, cooperative spirit, compassion, and so forth. It is by cultivating these qualities, or a genuine Tao of science, rather than by devising and enforcing a disciplinary protocol or by competing with its "natural enemies," such as religion, that Western and other science will become truly authoritative in a global environment. It is again Lao Tzu that describes, more fittingly than anyone I know, the principles that a future ecology of science, and of all other major human enterprises, should be based on and further developed:

The way is broad, reaching left as well as right.
The myriad creatures depend on it for life yet it claims no authority.
It accomplishes its task yet lays claim to no merit.
It clothes and feeds the myriad creatures yet lays no claim to being their master.
For ever free of desire, it can be called small; yet as it lays no claim to being master when the myriad creatures turn to it, it can be called great.
It is because it never attempts itself to be great that it succeeds in becoming great.

(*Tao Te Ching*, I.XXXIV)

Finally, from the perspective of a local–global ecology of science, the distinction between natural or physical sciences and human sciences, also known as a distinction between "hard" and "soft" sciences, becomes a specious one. All sciences are *human* sciences, because human beings produce them. Consequently, all branches of science, whether we call them natural, animal, or human, should serve the same purpose, that is, human development. In turn, such development will always be mindful of, and in harmony with, all other life on earth. In the end, the "hardest" of sciences, which has also been the most neglected by our mainstream scientists, is the science of being human.

III

Global Learning and Human Development

6

A Paradigmatic History of the University: Past, Present, and Future

In part II, I have outlined the principles of an ecology of science, oriented toward global intelligence. Since scientists receive their basic training in academic institutions, reforms of the ways in which we conduct scientific enterprises should begin with reforms of these institutions. Therefore, the project of an ecology of science would necessarily become part of a larger project to develop an ecology of education, equally based on an emergent ethics of global intelligence. In this third and last part of my study, I shall focus on current problems in higher education, particularly in the large North American research universities, and outline a number of preliminary steps that could be taken toward an ecology of global learning.

But, before describing the principles on which these various reform programs should be based, I would like to take a brief detour through the history of the idea of a university in the West so that we can better understand the educational choices that confront us in today's global circumstance. There are different theoretical models that can be employed in describing the history of a field, or of a branch of learning, or of an institution. The choice of any of these models will, however, result in different practical consequences not only for the way in which we look at this field or branch of learning or institution, but also for its future development.

In this chapter, I shall employ a number of such heuristic models to comment on the general direction of the university in the past and at present, as well as on its prospects for a paradigmatic shift in the future. The first model is a binary one and will attempt, at the most general level, to sketch the principles that I believe have informed the development of higher education since the creation of the university almost a thousand years ago. The second model will derive from the first one and will attempt to sketch, again very broadly, the disciplinary development of the university in the past century. The third and fourth models will briefly outline the general principles of a local–global intercultural university of the future.

I hope that these theoretical fictions, for which I claim no more cognitive status than that of working hypotheses, will nevertheless help us move toward creating substantially different models of advanced learning and research, appropriate for a global age. They will obviously need much refinement and fleshing out—a task that goes well beyond the purposes of the present book and could be assumed by future intercultural and cross-disciplinary teams of researchers in the history and theory of knowledge, grounded in an emergent ethics of global intelligence.

1 The University as Citadel and as Factory of Knowledge

It is widely claimed that in the past two decades or so the university has undergone sweeping changes, which will become even more sweeping in this new century, particularly under the impact of electronic information and communication technology. If we consider these changes from a cultural historical perspective, however, they may appear less radical, because from its inception the university has continuously been subject to conflictive impulses or forces that have pulled it in opposite directions. Very broadly speaking, one may call these conflictive impulses "aristocratic" or "elitist" on the one hand, and "democratic" or "popular" on the other.

In turn, the two conflicting impulses have laid the foundations for two academic paradigms that have emerged since the nineteenth century as dominant ones in Western academia and have been exported to other parts of the world as well: one paradigm organizes the university as a "citadel of knowledge" and corresponds to the aristocratic or elitist impulse; the other paradigm structures it as a "factory of knowledge" and corresponds to the democratic or popular impulse. Needless to say, these two paradigms cannot be found in a pure form in any specific modern university. On the contrary, they have been present all along, in one form or another, in most, if not all, historical institutions of higher learning, whose specificity may come precisely from various combinations and permutations of the basic elements of the two models. Finally, these two paradigms can be seen in a relationship of reciprocal causality, generating complex, amplifying feedback loops.

Although the history of the university can be traced as far back as Pythagoras, Eudoxus of Megara, Isocrates, Plato, and Aristotle, its modern avatars begin in the eleventh century at Bologna and in the twelfth century at Paris and Oxford. In the Middle Ages, universities began by training what today we would call "middle-class professionals" such as lawyers, physicians, and churchmen. In their first, medieval phase, therefore, universities can be seen as cultural forces that attempted to pull away from prevailing aristocratic values toward their middle-class counterparts; in other words, they can be seen as "democratizing" forces in a feudal society. A very brief overview of the beginnings and early development of the universities at Bologna, Paris, and Oxford reveals how academic institutions, from their very inception, had to contend for democratic autonomy against the various pressures exerted on them by the state, the church, and their surrounding communities.

In Bologna, academic teaching existed even before 1088, the conventional date for the foundation of the university, but it was at that point that it became free of ecclesiastical schools and struck out on its own. The university started with the study and teaching of canon and civil (Roman) law that became very important during the medieval power struggles between the church and state. This struggle ended in the creation of various European centralized states, headed by more or less absolute monarchies, with the basic support of the church. Law scholars at Bologna played a role in the establishment of the legitimacy of the Holy Roman Empire under Frederick I Barbarossa. In return, the emperor issued a *Constitutio Habita* (1158) according to which every school of learning within the empire must constitute a *societas* (society, association) of *socii* (students). The *socii* were taught by a *dominus* (master), to be paid with a *collectio* (collection) from the students. This collection was given to the *dominus* in the form of an *oblatio* (offering), because knowledge was considered a gift of God that could not be sold or bought.

Up to the sixteenth century, it was the students who ran the university at Bologna, on democratic principles, electing a rector from among their ranks. During the late twelfth century, they started organizing themselves in groups, called *universitates*. In Roman law *universitas* meant "corporation," and it is this original meaning that is relevant to the medieval university. (It was only later on that the term *universitas* began to be interpreted as referring to "universal" knowledge.) At Bologna, the *universitates* or corporations were formed by nonlocal students according to their place of origin, such as the Citramontanes (Italians from Lombardy, Tuscany, or Rome) and the Ultramontanes (non-Italians from beyond the Alps, such as the French, Spanish, Provençal, Catalan, English, German, Hungarian, Polish, etc.).

The academic corporations were primarily established in order to preserve student autonomy by coming to terms with the local commune or city-state. Bologna and other such Italian communes had wrestled their autonomy from Frederick Barbarossa and the Holy Roman Empire in the wake of the battle of Legnano (1176) and governed themselves according to what today one would call democratic principles. The citizens of Bologna found it to their advantage to preserve Barbarossa's law school that had meanwhile become famous all over Europe, but not without curtailing its freedoms, particularly the freedom of movement of its teachers. For this reason, the commune asked the latter to swear an oath that they would confine their teaching within the city walls. In return, the city started paying some of their teaching fees. One could thus say that in early thirteenth-century Bologna, the university had already attained its modern form in the interaction between students, teachers, and the nonacademic, local community that it began, willingly or unwillingly, to serve.

The Catholic Church also started involving itself in the power struggle between the commune and the university at Bologna, with the result that the teachers' oath of allegiance to the city was repealed, and papal authority to grant academic degrees was extended to the university. Finally, although the university began as a law school, divided into *collegia* of civil law and canon law, it diversified in the fourteenth century, adding a school of arts. This school, which comprised students of medicine, philosophy, arithmetic, astronomy, logic, rhetoric, and grammar, broke away from the school of law and formed its own *universitas* in 1316, electing another student-rector.

The university at Paris provided the other major model for the European university. This model was less democratic than the Italian one only in the sense that the *universitas* or corporation was headed by the masters, rather than by the students. Otherwise, the university at Paris was also born amidst power struggles for autonomy, in this case, from the jurisdiction of the French Catholic Church. This church had been running its own schools preparing church cadres since the time of Charlemagne. The university at Paris most likely came into existence in 1106 in connection with a famous controversy about "universals" between illustrious master logician, Pierre Abélard, and Guillaume de Champeaux, the *ecolâtre* (headmaster) of the church school under the authority of the Bishop of Paris.

Given the power of the bishop and his *ecolâtre*, Abélard deemed it wise to leave the Île de la Cité, where the church school was located, establishing himself on the left bank of the Seine, outside the bishop's jurisdiction. Here the university began to emerge around him and other masters and their disciples. Challenging the authority of the Bishop of Paris, as well as that of the king, groups of professors and students organized themselves in an autonomous corporation, the *universitas magistrorum et scholiarum parisiensium*, that is, the "university" or association of Parisian masters and pupils. The French king recognized them as such almost a century later, in 1200, and the pope did the same in 1231.

From Robert de Courçon's statutes for the University of Paris (1215), we learn that the university was especially strong in the liberal arts, divided into the *trivium* (grammar, rhetoric, and dialectic) and the *quadrivium* (arithmetic, geometry, astronomy, and music). Each master was to have jurisdiction over his scholars. The students were supposed not only to pursue their studies, but also to teach, after the age of twenty. The masters and students were to wear simple apparel, and their living habits and general demeanor were to be modest and self-restrained. (Thus, this statute indicates, indirectly, that the university was already affected by the ostentatious aristocratic manners prevalent at the French court.) During the thirteenth century, the arts, theology, law, and medicine constituted themselves in four separate *facultés* (faculties), with each electing its own dean. Whereas Bologna was best known for its legal studies, Paris was the leader in theological studies (and Montpellier in medical studies).

In turn, Oxford University, the youngest of the three, received a great boost in 1167, when Henry II banned English students from attending the University of Paris. By 1201 the university was headed by a *magister scholarum oxonie* (master of Oxonian pupils) who in 1214 was styled chancellor. In 1231, the masters were recognized as a *universitas* or corporation. During the thirteenth century, violent conflicts between the townspeople and the university contributed to separating the latter from the surrounding community by medieval walls and to the establishment of halls of residence.

From the beginning, Oxford, no less than its Continental counterparts, became a center of religious and political controversies and got involved in power struggles between the state and the church, which affected its autonomy. For example, in the fourteenth century, John Wyclif, Master of Balliol, challenged the papacy by advocating a vernacular version of the Bible. In 1530 Henry VIII forced the university to acquiesce in his divorce from Catherine of Aragon. During the Reformation, in the sixteenth century, three Anglican churchmen were tried and burnt at the stake in Oxford; and in the seventeenth century, the university sided with the royalist party during the Civil War.

During the Elizabethan age, the English university flourished by admitting a majority of students from the lower middle classes, such as sons of yeomen, tenants, cobblers, haberdashers, porters, and other artisans, as well as those of the middle class, such as sons of churchmen, schoolmasters, traders, and so forth, who received stipends from various public and private patrons, including the church. According to Lawrence Stone, between 1560 and 1640 Oxford and Cambridge received endowments for "some 500 new scholarships, intended to pay for the tuition and maintenance of poor boys" (Stone 1974, p. 21). The students at Oxford, or at least those who were seeking degrees in order to enter a profession, studied mostly to become clergymen and schoolmasters, rather than lawyers and physicians (who were largely apprenticed to and trained in the proximity of the law courts and the hospitals).

But there were a considerable number of gentlemen's sons who attended the university as well, not because they needed a degree, but because their parents wanted them to get a "liberal" education, which seemed to be the fashionable thing to do at the time. As Stone puts it, "distinct from its ancient role as a vocational training center for poor students aspiring to become clergymen, civil lawyers, physicians and schoolmasters, the university also came to serve as a place where the serious-minded country gentleman and the prospective common lawyer, secretary, or courtier could acquire further training in the classics, some useful experience in logic and rhetoric, and a smattering of Protestant theology, and could perhaps also study such things as history and modern languages" (ibid., p. 24).

In marked contrast to their democratic beginnings, however, universities soon turned learning into social privilege and prestige. Thereby they reaffirmed the very same elitist, aristocratic values they had challenged during their earlier history. Stone, for example, points out that after the Civil War and the Restoration, beginning with the seventeenth century, Oxford and Cambridge came to cater more and more to the upper classes, and less and less to the lower ones, so that aristocratic values and ideals prevailed over their middle-class counterparts. These trends are equally observable at Bologna and Paris, at least up to the French Revolution. Napoleon I, moreover, reversed the democratic trends, which had risen again in European universities during the French Revolution, toward more elitist ones.

Another important trend that came gradually to be reversed in the history of the European university was academic cosmopolitanism, expressed for example in the humanist idea of the "republic of letters." This aristocratic idea originated in Roman antiquity with Cicero's notion of *litterae publicae* (even though it could be traced back all the way to Plato's *Republic*) and was reformulated by Petrarch and others at the end of the Middle Ages. Although at first it was used as a weapon against the dogmatic, scholastic forms of knowledge favored by the medieval university, it was ultimately incorporated into the reformed university of the Renaissance and prevailed throughout "enlightened" Europe.[1] Academic cosmopolitanism led to a certain isomorphism of the European universities that, at least up to the nineteenth century, were fairly similar in terms of curriculum and institutional arrangements (Ashby 1966, pp. 3–5).

In the nineteenth century the universities became more and more differentiated: they increasingly identified with and aspired toward national ideals and practices, turning away from the cosmopolitan ones of the older European republic of letters (ibid., pp. 19–29). This trend became even more pronounced with the advent of the Industrial Revolution and positivistic science. Such developments brought further fragmentation to the cosmopolitan, albeit hierarchical, unity of knowledge (under the reign of philosophy–theology) on which the medieval university was based.

Even from this very brief historical sketch, one can see that, from the outset, the conflict between aristocratic and middle-class values in academia took the shape that can readily be recognized in today's universities. A full study would be needed to analyze the various binary oppositions generated by this conflict. Here I can only list a few of them, in shorthand: science and technology versus the liberal arts (a problem that is already present in Plato's *Republic* and *Laws* and Aristotle's *Politics*); corporatization versus freedom; town versus gown; utilitarian knowledge versus disinterested knowledge; disciplinarity versus transdisciplinarity; research versus teaching; instruction or training versus education. Of course, throughout the ages the university itself has often been perceived by outside, nonacademic innovators and reformers as an aristocratic citadel of (obsolete) knowledge, or a reactionary defender of the dominant cognitive and cultural paradigm. Thus, a full history of the university would also need to take into consideration the amplifying feedback loops between intramural and extramural knowledge production and transmission.

Many of the changes that the university underwent during the Industrial Revolution and that remain relevant today are reflected in John Henry Cardinal Newman's well-known collection of discourses, lectures, and essays on *The Idea of a University*, first published in 1852, but revised by the author in 1859 and then again in 1873.[2] Newman grapples with the same conflict between middle-class educational values, which he largely deplores, and aristocratic values, which he largely idealizes in his concept of higher education as a gentlemanly, disinterested pursuit toward intellectual excellence. Ironically, Newman—a clergyman—is invoking aristocratic or elitist values to defend the university against the dominant utilitarianism of his age. He thus defends it against the same practical values that had inspired the creation of academic institutions in the first place and had now become the victims of their own success.

Newman nostalgically harks back to the golden age of the medieval university, which he praises precisely for its cosmopolitan vision of universal knowledge and the cultivation of the mind: "The majestic vision of the Middle Age, which grew steadily to perfection in the course of centuries, the University of Paris, or Bologna, or Oxford, has almost gone out in night. A philosophical comprehensiveness, an orderly expansiveness, an elastic constructiveness, men have lost them, and cannot make out why. This is why: because they have lost the idea of unity: because they cut off the head of a living thing, and think it is perfect, all but the head" (Newman 1947, p. 393).

There is a rather transparent allusion in this passage to the literal and symbolical beheading of the English and French monarchs in the two previous centuries at the hands of "democratic" social forces. Symbolically, this violent act meant the severance of the king's subjects (his body) from the divine fountainhead (the king's head) and, consequently, the scotching of the medieval Great Chain of Being, including the hierarchical unity of knowledge that it involved. No wonder, then, Newman implies, that the university has equally lost its sense of intellectual unity and has become fair game for middle-class, godless and headless, utilitarian pursuits.[3]

Newman blames John Locke, among others, for the English utilitarian idea of a university (ibid., pp. 140–141). He notes that some "great men" such as Locke "insist that education should be confined to some particular and narrow end, and should issue in some definite work, which can be weighed and measured. They argue as if every thing, as well as every person, had its price; and that where there has been a great outlay, they have a right to expect a return in kind. This they call making Education and Instruction 'useful', and 'Utility' becomes their watchword" (ibid., pp. 135–136).

By contrast, for Newman, academia should separate the pursuit of truth from humankind's "necessary cares." It should pursue knowledge and intellectual excellence for their own sake and it should be "the high protecting power of all knowledge and science, of fact and principle, of inquiry and discovery, of experiment and speculation" (ibid., p. 66). For Newman, the university is "a place of *teaching* universal *knowledge*" (ibid., p. xxvii; emphasis in the original). This means that "its object is, on the one hand, intellectual, not moral; and, on the other, that it is the diffusion and extension of knowledge rather than the advancement. If its object were scientific and philosophical discovery, I do not see why a University should have students; if religious training, I do not see how it can be a seat of literature and science" (ibid.). The function of the university, therefore, is not to advance knowledge and/or morality, but only to preserve and protect them in their present form, just as a citadel protects its civilized inhabitants from the onslaught of barbarian outsiders.

Newman also draws a distinction between education and instruction: the end of the former is "to be philosophical," the end of the latter is "to be mechanical." Education "rises towards general ideas," whereas instruction is "exhausted upon what is particular and external" (ibid., p. 99). In other words, instruction is utilitarian, preparing students for a profession and therefore its proper place is not the university, but a professional school (ibid., pp. 100–101). By contrast, education is an end in itself, preparing students towards social life as a whole. It is "liberal" in the sense of forming a "habit of mind" that would "last through life, of which the attributes are freedom, equitableness, calmness, moderation, and wisdom" (ibid., p. 90). One may note that these also became the virtues of the idealized Victorian gentleman, as promoted by the official ideology of the British Empire at its zenith. Finally, a university is "an Alma Mater, knowing her children one by one, not a foundry, or a mint, or a tread mill" (ibid., p. 128).

In Newman's essay we can therefore discern the two academic models that have been in conflict ever since the foundation of the university: the academy as an elitist citadel of knowledge and the academy as a more or less democratic factory of knowledge. Newman is obviously in favor of the first model, although he makes some concessions to the second one as well. In addition to the metaphor of the factory, Newman employs other commercial metaphors that are equally appropriate for the "democratic" university. For example, he compares it to the "pantechnicon" or the department store, which originated in the Industrial Age, on the model of the Eastern bazaar.

According to Newman, defenders of the utilitarian university "consider it a sort of bazaar, or pantechnicon, in which wares of all kinds are heaped together for sale in stalls independent of each other; and that to save the purchasers the trouble of running about from shop to shop" (ibid., p. 391). He further compares it to a lower-class "hotel or lodging house where all professions and classes are at liberty to congregate, varying, however, according to the season, each of them strange to each, and about its own work or pleasure" (ibid., pp. 391–392). By contrast, the "right" kind of university is more like a genteel Victorian mansion: "if we would rightly deem of it, a University is the home, it is the mansion-house, of the goodly family of the Sciences, sisters all, and sisterly in their mutual dispositions" (ibid., p. 392).

All of these metaphors, revealing the conflict between the democratic and the aristocratic model of higher education, will hardly lose their relevance, but will become even more appropriate throughout the twentieth century. They obviously apply to the North American university as well. In the past few decades, especially in the United States (but also in Canada), the dominant academic paradigm has been that of the "factory" or the "department store" of knowledge, although the aristocratic tendencies have by no means been entirely subdued. For instance, they can be recognized in the principle of "academic excellence" (equally present in Newman) that is currently assuming more and more distorted, bureaucratic forms in the large North American research universities.[4] They can also be recognized in such U.S. academic institutions as the Ivy League and the small liberal arts colleges. They are no less present in the academic star system and its counterpart, the academic slave-labor system, with its own kind of academic sweatshops, where a growing number of part-time instructors and teaching or research assistants teach a large number of courses or carry out mechanical, laboratory research tasks for minimal wages.

In turn, the liberal arts (the term "liberal" in the phrase can be traced back not just to Newman, but all the way to Aristotle's view of scholastic knowledge as a form of aristocratic play, unfit for slaves), or the humanities as a whole are put in the unenviable position of defending elitist values against an overwhelmingly utilitarian, middle-class tendency in contemporary academia that has gained even more momentum since Cardinal Newman's day. In this regard, even the current North American academic trends of liberal "political correctness," which at first blush appear to be of a democratizing nature, are paradoxically the efforts of an elite to preserve its privileges by accepting a limited number of token minorities among its ranks. For example, in the field of literary studies, it is precisely the critical elite (teaching at elitist universities) that attempts to supplement the traditional literary canon with a few works that would satisfy the demands of the more vocal ethnic minority groups in North American academia.[5]

One can also point out that universities have in the past century developed their own hierarchical, aristocratic culture, which has survived in academia to an extent no longer possible in other social milieus. At the top of the pyramid, one can find the Ivy League schools in the United States, "Oxbridge" in England, and the so-called Grandes Ecoles in France. Especially in North America, one can also find a relatively small number of second- and third-tier research, public and private universities in the middle, and a large number of four- and two-year community colleges at the bottom of the pyramid. The schools at the top have been entrusted with the education of the new, meritocratic elites that make most of the important economic and political decisions in the Western world.

2 The University as Disciplinary Institution

The previous point brings us to the disciplinary structures of the contemporary university that continue, to some extent, to reflect an uneasy, yet archaic and well-entrenched, cohabitation between aristocratic, hierarchical values and democratic, egalitarian ones. These structures have paradoxically survived the great social upheavals of the past century with only a few administrative reforms that have in effect managed to increase its disciplinary (but not necessarily its educational) efficiency. For example, North American academia has generally moved from the relatively benign, paternalistic administrations of the small, liberal arts colleges and universities that were still common in many parts of North America during the first half of the twentieth century to the bureaucratic mammoth machines of the contemporary research and "mega" universities. From the standpoint of global intelligence, one may call these bureaucratic machines inefficient, but from a disciplinary standpoint they are efficient enough in relation to their main objective, which is, like that of any other power machine, self-enhancement and self-preservation.

North American contemporary academia (although European and other academic worlds are cases in point as well, albeit in different ways that could be objects of comparative analysis for our intercultural research teams) is living proof that Michel Foucault's thesis that power has effected a historical transition, in the so-called developed societies, from social disciplinary structures to ones of social control is only partially plausible. It is true that Foucault intended his model for application in a French cultural context, but even there it does not quite work in the form he presents it. Foucault's analyses in *Surveiller et punir* (1975) and other books attempt to reveal the ways in which power organizes all social life and institutions, including the university. To this purpose, Foucault makes a theoretical distinction between "disciplinary society" and "the society of control," which he then presents diachronically: he postulates a historical transition from one to the other.

Foucault places the emergence of his disciplinary society in the classical age of French culture or the Age of Reason, and describes it as a diffuse network of social mechanisms that regulate economic production, social practices, customs, habits, and all the other elements of socioeconomic and cultural life. Disciplinary institutions such as the prison, factory, workshop, market, army, hospital, courthouse, school, and so forth ensure that all members of society obey the same rules through powerful mechanisms of inclusion and exclusion. Reason itself is pressed into the service of disciplinary power, being called to elaborate and justify a disciplinary logic and discourse in general, on the basis of the same inclusive or exclusive mechanisms. According to Foucault, therefore, disciplinary power rules by establishing strict limits, not only to social practices but also to thought, which is monitored as closely as any other activity, for example, through the institution of censorship.

By contrast, the society of control, the advent of which Foucault locates historically somewhere between modernism and postmodernism, employs ruling mechanisms that are more democratic and, therefore, more insidious, because they have now become internalized throughout the bodies and minds of the citizens. The society of control does not operate so much through visible disciplinary mechanisms as through vast, invisible networks. Foucault also describes this new paradigm as biopower, because it entirely engulfs and regulates social and private life from inside an individual's body and consciousness, indeed becomes a lifestyle for each and every member of society. Information technology and the new media, among other postmodern inventions, are precisely such networked, all-pervasive, internal mechanisms of control (cf. the similar, post-Marxist views in part I above).

One should point out, however, that Foucault's theoretical models might work much better synchronically, rather than diachronically:

power has always combined discipline with control in any society that organizes itself according to its principles. A society of control cannot operate without a disciplinary paradigm, any more than a disciplinary society can function without a control system. Disciplinary and (self-) controlling methods may, of course, differ from age to age and may certainly include the postmodern information technologies and the new media. They may also appear as more or less visible or conspicuous in various societies.

Foucault's work is valuable insofar as it develops Nietzsche's intuition that, for a certain kind of human mentality, power is all that is and has always been. In other words, in Foucault's description of the society of control, power finally reveals itself as all-pervasive, not just as exterior domination or social control. But, this revelation occurs not because power has gradually become more internalized—it has always been both internalized and externalized, both inside and outside the individual, that is, it has always been a lifestyle or biopower. Rather, the revelation becomes possible, paradoxically, because the more socially visible, hierarchical forms of power have been gradually replaced by less visible, more democratic forms under the constant pressure for more power sharing by more and more members of a particular society.

The disciplinary mechanisms of social inclusion and exclusion have not disappeared, but have moved *out of sight* as it were, even though they are an implicit guarantee of any form of control. In the "developed" societies, their visible, outside forms have become more and more limited, first to women, then to minorities, then to noncitizens or foreigners, and then to the so-called fringe elements of society such as criminals, drug addicts, sexual "deviants," "illegal" immigrants, and so on. On the other hand, the more the individual members of a society clamor for and get a larger share of power (be it only nominally), the more power becomes visible as emanating from inside, rather than from outside them. Joanna Macy's and other social activists' so-called empowerment is precisely this kind of internalized power. Indeed, the inside/outside dichotomy itself is nothing but a power strategy that becomes revealed as such in postmodernity.

Once this strategy becomes fully exposed within a certain community, power exports its more visible disciplinary mechanisms to other societies, be it through colonialism, imperialism, or economic and political clientelism (which will sooner or later come back to haunt the exporters). It is for this reason that Foucault's strategy of opposing and undermining "centers of power" wherever they might be found can only have the contrary effect of multiplying these centers until it becomes apparent that each and every individual from a particular culture is, at least theoretically, such a center of power. It is also for this reason that Michael Hardt and Antonio Negri's idea, in their book *Empire* (2000), of a multitude that will struggle against and undermine the empire from the inside is no more than a power-enhancing fantasy.

Hardt and Negri are more credible when, for example, they acknowledge, if only indirectly, that capital cannot function without the cooperation of labor, which previously was largely physical and now is increasingly intellectual. Labor, whether physical or intellectual, has always been engaged in a relationship of mutual causality with capital and is one of its conditions of possibility, precisely because of the promise of reversibility inherent in any power dynamic. It is what capitalist wisdom calls equality of (socioeconomic) opportunity, rather than equality *tout court*. It is in fact this eternal promise of reversibility that Hardt and Negri dangle before the new intellectual proletariat of the information age. But why should the new proletariat listen to this particular siren's call, when capital has now become even more mobile and fluid and therefore even more reversible or, shall we say, nomadic? Within "developed" societies, struggling against capital appears more and more like struggling against oneself. (Jacob and his angel might be another good metaphor for the reciprocal causality of the relationship between capital and labor.)

But, with the advent of modernism and postmodernism, the visible forms of disciplinary society—which in an academic context may be called disciplinary culture—have certainly not disappeared from the university or from the ways in which we carry out and utilize scientific research. On the contrary, academic and scientific disciplinary cultures coexist side by side with the culture of control described by Foucault. This odd time loop that is the university has even been able, at least for the moment, to withstand (rather than be transformed by) the vast expansions of knowledge in certain scientific fields. These expansions, given the obdurate, disciplinary nature of the academic world, have inevitably led to further divisions, rather than integration, of knowledge.

Yet, the human division of knowledge, no less than the human division of labor, is not by itself a phenomenon of human alienation. From the perspective of an emergent ethics of global intelligence, the problem with the current ways in which we acquire, transmit, and utilize knowledge does not come from academic or any other type of specialization or nonspecialization, but from the disciplinary mentality that preoccupies these ways—and I mean "pre-occupation" in a military sense as well. Just as one should beware of confusing locality with localism or provincialism, one should beware of confusing academic or any other specialized form of knowledge with what one may call academic (and nonacademic) disciplinarianism, with its paranoid turf-guarding and bureaucratic methods.

Conversely, just as one should beware of political globalitarianism, one should beware of its cognitive counterpart, academic or scientific generalism, with its grand theories and imperialist claims that attempt to marshal and regiment all knowledge into one Procrustean framework (we recall the example of Wilson's sociobiology or Weinberg's elementary particle physics in chapter 3 above). A generalist with a strong theory of "everything" can be as much of a disciplinarian as a specialist or an expert: what they have in common is that both of them claim to know more and better than anyone else. In this sense, one may also speak of the macro- and micrototalitarianism of the academic (and nonacademic) disciplinary mentality, which translate into its bureaucratic administrative structure as well.

One should therefore not neglect the fact that the emergence of positivist ways of thinking in the late nineteenth century may in the end have contributed to the bureaucratization of academia and vice versa, that is, that they are in a relationship of reciprocal causality as well. Positivist thinking may thus have contributed to the rise of positivistic scientific culture, the explosion of industrial technology, the proliferation of narrowly conceived fields of specialization, and the corresponding fragmentation of traditional disciplines, the formation of a technocratic mentality, and the establishment of a managerial educational model. It has also contributed to the twentieth century's polarizing tendencies with regard not only to political systems and ideologies but also to human thought, emotion, and behavior in general. In this sense, they have also amplified tendencies to devaluate "liberal," humanistic education by promoting sectarianism and dogmatism over a free play of ideas and an open debate of issues of common human concern.

We have seen that recent nonlinear thinking in the sciences has called into question the primacy of positivistic, deterministic, and technocratic worldviews. Certain strands of systems and information theory, chaos theory, models of emergence and complexity, polysystems theory, and so forth point to a nondeterministic and nontotalizing world outlook that goes beyond conflictive polarities and is mindful of both local and global conditions. Many of the ground principles of this world outlook have already been articulated in humanistic thought for centuries, and it is at the level of theory that one can again elaborate a common discourse between the humanities and the sciences. Yet, as we have equally seen in previous chapters, we are very far from such an integrative discourse and even farther away from abandoning the disciplinary practices that have brought us to the present state of affairs in the first place.

The medieval hierarchies of the arts and sciences (the trivium and the quadrivium), with theology as undisputed queen, shifted slightly in the Age of Reason, when the place of theology was taken by philosophy. Yet, they have hardly disappeared in the modern university. On the contrary, they have instead greatly multiplied and flourished. In a sense, these hierarchies have come full circle since medieval times, according to the dynamic of center and margin that I have discussed in relation to liminality: natural science, which was once banished by theology–religion to the margins of its empire, has now replaced its old foe as the uncontested queen of the modern, bureaucratized *universitas*, even though for different, often venal, reasons. We have seen, in chapter 3 above, that Thomas Kuhn, as a historian and philosopher of science, and E. O. Wilson and Steven Weinberg, as practicing scientists, indirectly confirm this judgment. We recall, for example, Weinberg's arguments for the construction of an exorbitantly priced Super Collider in Texas, corroborating Wilson's observation that the "eyes of most leading scientists, alas, are fixed on the gold" (Wilson 1998, p. 41).

We also recall Kuhn's remark that scientific training involves "a narrow and rigid education, probably more so than any except perhaps in orthodox theology" (Kuhn 1970 [1962], p. 166). So Kuhn's term "disciplinary matrix," which he applies to normal or mainstream science, can be understood in both senses of "disciplinary" that I have discussed here: (1) a scientist's professional "discipline" or specialized field of knowledge and (2) the (self-)controlling structure of scientific practice itself.

As Kuhn points out, individual members of a particular scientific community are hardly expected to deviate from the disciplinary matrix–paradigm that is currently in force and are often sanctioned if they do so. Therefore, they internalize its rules and become particularly blind to any scientific discovery that might infringe on such rules or invalidate their disciplinary paradigm. In other words, they are perfect illustrations of the counterproductive, amplifying feedback cycles that operate within disciplinary frameworks, where disciplinary structure and (self-)control mutually enforce each other. Kuhn finally notes, we recall, that scientific communities are "peculiar kinds of democracies" within the larger democratic framework of our Western societies.

All of the preceding observations are valid not only for our scientific communities, but also for our academic communities in general: they also form "peculiar kinds of democracies" within our democratic societies at large. I believe that most of us who teach and conduct research in the humanities in North American universities are familiar with the various critical fashions that prevail in our fields and that largely determine student admissions to graduate programs, hiring junior faculty members, and granting them promotion and tenure. These critical fashions constitute a sort of "second-order" disciplinary paradigm, specific to a particular branch of learning and nested within the larger, or primary, bureaucratic paradigm. This latter paradigm is presumably based on the impersonal, "objective," principles of so-called excellence or professionalism, regulating all academic activities.

The two disciplinary paradigms are equally engaged in amplifying feedback loops that mutually enforce each other. When they do clash occasionally, it is mostly because the second-order paradigm shifts more often than the primary one and therefore it strives to be accepted by the latter. A graduate student in the humanities (although this phenomenon, I suspect, extends to other academic disciplines) must belong to one of these critical fashions and work within the accepted frameworks of both paradigms, if she or he wishes to have a successful academic career in the United States.

To give a brief example, again from the humanities, the prevailing critical fashion or research paradigm within the North American humanistic disciplines of the 1940s and 1950s was new criticism, which was replaced by structuralism, which in turn was replaced by poststructuralism, which has currently been replaced by postmodernism with its shifting alliances of postcolonial, Marxist, deconstructionist, gender, cultural studies, and other critical trends. Of course, prevailing second-order disciplinary paradigms overlap and contend among themselves at various universities. But, the Ivy League or Ivy-League-style schools (e.g., on the U.S. West Coast) are as a rule the tacitly accepted paragons that all the rest will follow (just as Milan, Paris, London, or New York are leading paragons in the global fashion industry). A graduate student must therefore choose to work within one of the latest critical paradigms and at one of these "first-order" schools, if he or she wants to land the most prestigious job and/or the most prestigious research grant. Otherwise, she would either have to leave the profession or content herself with a modest position at a third- or fourth-rate university—a hierarchy that is, again, largely determined according to the measuring scale of the bureaucratic/meritocratic principle of excellence.

I should also emphasize that interdisciplinarity, at least in its present form, is not a way out of the disciplinary academic paradigm, but a way of reinforcing it. Interdisciplinarity became the order of the day in North American universities in the 1960s and since then it has been exported to universities in other parts of the world. It might initially have had something to do with the temporary liberalization of the disciplinary university under the impact of student movements in Western Europe and the United States. But, since then, it has been promoted primarily by upper administrations and not, as one might have expected, by academic departments or by students: during the last quarter of the twentieth century, university administrations found it increasingly hard to deal institutionally, under the disciplinary paradigm, with the explosion of knowledge led by the physical and life sciences and, more to the point, with its potentially profitable applications in the new economy.

A handful of "soft," but venerable, traditional disciplines such as literary studies also jumped on the interdisciplinary bandwagon (even though the initial bureaucratic call was most likely not addressed to them, but to the "hard" sciences). They formed departments or programs of general literature, comparative literature, modern thought and literature, cultural studies, women's studies, and so on. Such soft disciplines have had no other recourse but to go interdisciplinary precisely in order to be able, paradoxically, to cling to their disciplinary identities and academic departments, even though this strategy has so far yielded only mixed results. Philosophy, history, anthropology, and other ambiguous disciplines hovering between the humanities and the social and physical sciences have also employed this survival strategy, with equally mixed results, with the possible exception of environmental studies. None of these fields has recovered from the onslaught of logical and scientific positivism and academic bureaucratization, however, and none has become genuinely cross-disciplinary, let alone cross-cultural.

With the advent of information technology and other technosciences, the pressure of interdisciplinarity, that is, the disciplinary attempt at the integration of knowledge necessary for further scientific and technological advancement and, more significantly (from a bureaucratic standpoint), for increased financial profit, has grown exponentially. This pressure might eventually explode the very same academic disciplinary structures that made interdisciplinarity possible in the first place. In other words, the growing gap between the disciplinary institutional arrangements and the type of cross-disciplinary and cross-cultural knowledge needed for our networked, globalized economies might eventually force the university to start moving from a predominantly disciplinary culture toward a "postmodern" culture of control. It is doubtful, however, that the university will turn away from its bureaucratic disciplinary mentality any time soon, unless there is also a clear desire for extensive educational reform not only on the part of a large number of academics, but also on the part of the civil societies whom the university is supposed to serve.

3 The University without Walls or without (Red) Brick and Mortar

With the advent of the information age, we also see some elements of the two paradigms, the university as citadel of knowledge and the university as factory of knowledge, slowly emerging into what might conceivably become a third paradigm: the university without walls or without (red) brick and mortar. The metaphors of "citadel" and "factory" both convey the disciplinary nature of the university. In the first case, there is the aristocratic idea of knowledge as power—be it under the gentlemanly guise of a free, disinterested, and leisurely intellectual pursuit—that must remain the privilege of a few and therefore must be jealously guarded and defended from the assault and penetration of the masses—here one may think of such hoary British academic establishments as Eton, Oxford, and Cambridge before World War I.

In the second case, one may develop the factory analogy by describing specific academic forms of Fordism, Taylorism, and their contemporary "neo-" and "post-" variants, as well as other bureaucratic rationalizations of intellectual labor, typical of academic mass production. One should also not forget the academic sweatshop as a source of cheap intellectual labor, mentioned in the preceding subsection. In turn, under today's neoliberal and bureaucratic academic paradigm, a sizable part of the university, concerned particularly with the technosciences, has become a relatively cheap source of intellectual labor for the corporate world, or what one may call a "gilded sweatshop."

A university without walls or without (red) brick and mortar, on the other hand, should metaphorically connote the removal of the disciplinary barriers that have been erected everywhere by the traditional *universitas*. But it should also connote the e-commerce vocabulary that describes selling and buying a wide range of goods through the Internet, rather than in conventional stores, built out of brick and mortar, that is, the postmodern, electronic version of Newman's pantechnicon or bazaar.

In the electronic information age, universities may become one of the service industries, and in many parts of the United States they are in fact already looked upon and treated as such. In the words of Robert Reich (2001), they offer "symbolic-analytical services." As part of the service industry, universities supposedly produce and move around intellectual or immaterial commodities that belong to a so-called cultural capital (Pierre Bourdieu's phrase). Thus, universities, together with the new media, are often seen as partly belonging in the information industry and partly in the entertainment industry (in the low-paying range of the latter, according to the quip of a colleague's father). In keeping with these e-commerce trends, there are a large number of virtual universities and professional schools that have mushroomed both in the United

States and in other parts of the world and that more or less share the e-commerce philosophy. At the same time, brick-and-mortar universities equally seek to offer some of their courses and programs online and have even formed transatlantic partnerships to do so.

"University without walls" may connote, in addition to or in combination with the virtual university concept, a large number of centers or institutes of continuing education that are catering to a rapidly growing, nontraditional student population of all ages and walks of life. They seem more promising than the virtual university with respect to creating and disseminating new learning structures and methods that depart from the strictly disciplinary ones. In the case of the virtual university, for example, disciplinary courses have oftentimes been simply exported to virtual space without any modification or much regard for the knowledge content and learning potentialities of the electronic medium.

In this respect, the virtual university has duplicated and has made more efficient the older form of the "university by correspondence" that employs similar, long-distance, disciplinary learning methods, conveyed through "snail mail." Far from inhibiting the disciplinary instincts and habits of the brick-and-mortar university, the electronic medium has offered the latter the possibility of rapidly replicating and disseminating them in virtual space ad infinitum—a sort of Fordism and Taylorism online. Consequently, virtual universities have as yet not posed a present and clear threat to the disciplinary culture of the brick-and-mortar university, although they have moved aggressively into the education "market" and are beginning to cut into its profit margins.

On the other hand, the centers or institutes for continuing education could become the base for a different kind of academic institution, assuming they could detach themselves from the current bureaucratic power structures of the brick-and-mortar university. That they pose a real threat to the latter is witnessed by the fact that many academic departments reclaim courses in their disciplines that are taught under continuing education and clamor for full control in deciding the content of and hiring the teaching staff for these courses. These academic departments base their demands on their "legitimate concern" for the quality of instruction and content that should conform to the disciplinary standards and regulations currently in force within each discipline at a given university.

"Quality control" is the watchword of bureaucratic academia, as it has always been in our brick-and-mortar factories and stores, but it refers less to enhancing or maintaining the high quality of education (however "high quality" might be defined), than to keeping it within strictly prescribed disciplinary confines. Quality control, whether in the student requirements for a degree or the faculty requirements for promotion and tenure, ensures that the university remains a highly structured, regulated, and controlled, disciplinary environment.

Continuing education, through its informal learning environments, flexible teaching schedules, cross-disciplinary and extracurricular offerings, and nondegree course requirements and credits, as well as through its widely diverse student populations and closer proximity to the community, may eventually contribute to the reform of the university as a "halfway house of knowledge." This metaphor preserves the disciplinary connotations of the university in the law enforcement concept of a "halfway house," that is, a place situated halfway between prison and home, where exconvicts or drug addicts are socially reeducated or "redeemed," before being released back into the community.

The metaphor also preserves the connotation of e-commerce, for which a "halfway house" is a warehouse, situated halfway between the place of production and the buyer's domicile, where commodities are stored before being shipped out to the consumer. For contemporary e-commerce, for example, Ireland is a preferred location where such halfway storage- or warehouses are built, precisely because this island is geographically situated at the crossways between Europe, Northern Africa, and North America. Thus, the university as a halfway house may equally convey a third idea, that of a node or a place of intersection in an exchange or communication network, in this case an exchange or communication network of intellectual or immaterial commodities.

Other promising nodes of intersection between the university and the outside community could be the so-called university extensions, which were created in the nineteenth-century North American land grant colleges in response to the practical (largely agricultural) needs of local communities. Land grant colleges came about through an act of the U.S. Congress in 1862. By this piece of legislation, each state put aside 30,000 acres of land to form a perpetual fund for the endowment and maintenance of "at least one college where the leading object shall be without excluding other scientific or classical studies, and including military tactics, to teach such branches of learning as are related to agriculture and the mechanic arts, in such manner as the legislatures of the States may respectively prescribe, in order to promote the liberal and practical education of the industrial classes in the several pursuits and professions in life."[6]

Although this "democratization" of the traditional liberal arts college obviously has its advantages and disadvantages, originally it had the salutary effect of orienting higher education toward the needs of the community. As we can see from the text of the act, however, the needs of the community were far from being perceived entirely in practical terms, the emphasis being placed equally on both "liberal" and "practical" education. ("Military tactics," on the other hand, is a transparent reference and concession to the needs of the Civil War that was very much on the minds of the American legislators in 1862.)

In the course of time, however, the "practical" took over the "liberal," and *education toward* "the pursuits and professions in life" was replaced by *training for* such pursuits and professions. As Eric Ashby puts it, gradually "the idea took shape that a university should be a sort of intellectual department store offering courses in an extraordinary range of subjects, from how to dance to how to bury the dead" (Ashby 1966, p. 17). Note Ashby's commercial metaphor, borrowed from Newman, which corroborates my description of the brick-and-mortar university. Even so, the university extension could become an important and appropriate link between the academic and the nonacademic community, once it is emancipated from the bottom line utilitarian concerns of bureaucratic academia and once academic and nonacademic information and knowledge start flowing freely both ways.

The role of the university, then, would be not only to generate new knowledge, to debate and to exchange ideas, but also to facilitate their free flow both inside and outside the academic communities throughout the world. In this sense, the university as "halfway house" would finally convey the idea of liminality that I have discussed in previous chapters. The paradigm of the university as "halfway house" will go beyond the other three (university as citadel, as factory, and without walls) only if it can, under the impact of globalization, transmute and transfigure their basic principles and practices into a different, global reference frame, in which cooperative, rather than competitive relationships of reciprocal causality may obtain. In other words, the university as halfway house will become a genuinely new paradigm only if it can fulfill its liminal vocation and become a local–global institution, oriented toward global intelligence.

On the other hand, I cannot emphasize enough that one should by no means underestimate the obdurate, well-entrenched, bureaucratic conservatism of the contemporary academic culture, nor should one extol indiscriminately any kind of globalizing trend. From the standpoint of the university and other bureaucracies, globalization is simply one more bandwagon to jump on and will essentially not change their main objective: to preserve and enhance their own power by whatever means. It is for this reason that university bureaucracies (as well as state or suprastate bureaucracies) have subscribed to the neoliberal financial, economic, and political program of globalization, with its call for the universal adoption of neoliberal business models in all domains of human activity, including education and culture.

Thus, well-meaning and in many ways useful studies such as those included in *Universities and Globalization: Critical Perspectives* (1998), edited by Janice Currie and Janice Newson, largely ignore the fact that university and state bureaucracies have been in place long before (neoliberal versions of) globalization came into vogue and will possibly survive this particular wave as well. It would be obscuring the real issues, if we linked too closely the advent of the bureaucratic university with the advent of globalization, as some contributors to the volume seem to do. (This, of course, does not mean that we should ignore the obvious amplifying feedback loops between the bureaucratic academic paradigm and certain neoliberal globalizing trends, as these contributors rightly point out.)

It may seem paradoxical that the bureaucratic university would adopt the neoliberal program, which would appear, at least at first sight, to be profoundly inimical to it, with its calls for fewer bureaucratic controls and regulations and for more entrepreneurial, laissez-faire policies. It is plain to see, however, that the university bureaucracies (as well as state bureaucracies and corporate elites in general) are applying these kinds of neoliberal policies to everyone else but themselves. The phenomenon of exemptionalism, characteristic of any power-oriented mentality, is fully operative in this reference frame as well. Such corporate slogans as "teamwork," "productivity," "quality control," "transparency," and "accountability" are meant to function in one direction only, from top down, and not vice versa.

Furthermore, upper bureaucratic and corporate ranks have not decreased, but have steadily increased with the advent of neoliberal globalizing trends, despite repeated, well-meaning efforts of "reinventing" government. In this regard, the academic unionists' calls for resistance against globalization (read: neoliberal versions of it) do not affect in the least the university and the state bureaucracies. In fact, they are given an extra boost, when these unionists advocate, nostalgically (and ironically), a return to previous state regulation and control of neoliberal business practices, which presumably served public, rather than private interests. In the end, however, no matter who wins out, whether it is the neoliberal or the state and suprastate control advocates, the general public will lose, and the university bureaucracies, no less than their state and corporate counterparts, will continue to grow and prosper.

Most of the contributors to *Universities and Globalization* largely fail to realize that the enemy (or angel?) they are grappling with is not outside, be it globalization, the neoliberal creed, or even the bureaucratic and corporate structures, but inside them. This enemy or angel effectively controls even their well-meaning academic labor unions that do little more, through their "resistance," than reinforce the power structures they attempt to oppose. The authors certainly seem right to me in pointing out that neoliberal globalizing trends are neither irreversible "fate" nor the only possible paths to globalization and that, if they are not abandoned or at least substantially tempered, they will seriously inhibit human development, including its socioeconomic aspects. But I cannot believe that a lasting way of ensuring continuing human development would be to revert to older bureaucratic practices of state or suprastate control and regulation, be it in the name of the "public good."

Nor would it be of much help—as I argued at the end of part I—to introduce throughout the world the Western model of liberal democratic society, with its allegedly universal democratic values that some of the contributors to the volume, including Janice Newson, seem to advocate. Of course, this societal model might be appropriate for certain communities—and ultimately it is up to those communities to embrace it or not (and hardly up to the Western democracies to impose it by force). But it would be much more beneficial, at least from the standpoint of an emergent ethics of global intelligence, if we could be persuaded collectively to turn away from *kratos* (power) altogether. This word is omnipresent in our vocabulary, from the Greek etymological root of "democracy" to a plethora of terms describing our various forms of sociopolitical arrangements, such as aristocracy, bureaucracy, meritocracy, plutocracy, and so forth. Once we organize human relations on grounds other than power, new cultural, socioeconomic, and political forms, which may at present appear as inconceivable or utopian to our local mentalities, will undoubtedly emerge.

Janice Newson comes closer to a viable position, from the standpoint of a local–global theory and practice oriented toward global intelligence, when she describes the unexpected side effects of the strike that her academic labor union staged against the upper administration at the University of York in Canada. While on the picket lines (which can also be regarded as a form of liminality), colleagues from various departments and "disciplines" started for the first time talking to each other. During these friendly conversations, they discovered common research and other concerns and developed an interest in working together on various "interdisciplinary" projects. It is this kind of amicable dialogue and collective desire to engage in transdisciplinary and cross-cultural research and other projects for the benefit of all communities (whether academic or nonacademic) that will eventually advance human development throughout the planet. In turn, the project of a liminal university would require going beyond both the contemporary bureaucratic paradigm and the academic unionist or "democratic" one, but also beyond the disciplinary paradigm as a whole, in all its academic and nonacademic aspects.

4 The Future Role of the University

It might be instructive to place the model of a liminal university and its local–global networks envisioned here in a comparative historical perspective and juxtapose it to the liminal model of the monastery of earlier ages, at least as imagined by mystics and visionaries, such as Gautama Buddha and his disciples in the East, or Francis of Assisi, Teresa of Avila, and Juan de la Cruz in the West. In fact, these visionaries sought to reform their traditional religious institutions by reaffirming their liminal nature. (We recall, for example, that the Buddha retreated to the Himalayas in order to meditate on questions of government and power.)

Monasteries were originally conceived and built as liminal sites located—metaphorically, but often also geographically—halfway between the desert and human habitation, as well as halfway between the City of Man and the City of God. On this liminal ground, small communities of men or women opted out of the world of power in order to live a quiet but productive life of peace, meditation, and prayer. They meant to help each other and those from outside to develop their moral and spiritual dimensions. They had also gradually built a network of such liminal sites throughout the known world and had established steady flows of communication and knowledge exchanges between them. They effectively communicated, not through the Internet, but through long and perilous voyages and pilgrimages, as well as through extensive circulation of manuscripts and edifying oral narratives.

The ancient monastery, however, soon strayed from its original mission of preparing a different world for itself and others, while still in and of this world. Instead, it simply started replicating the power structures it had supposedly left behind. Thus, its liminal vocation became obscured. Even worse, in the case of Christianity, its networks became divided and hierarchical, with one center of command and control located in papal Rome and another, in Eastern Orthodox Constantinople. The monastery also started accumulating great material (rather than spiritual) wealth. Consequently, it often became the scene of power struggles among rival church and other dignitaries, rather than a place of contemplation and of knowledge production and exchange, to the benefit of all humanity. Repeated efforts on the part of mystics and other church reformers to reaffirm the liminal nature of the monastery and the church in general (which was equally supposed to mediate between the City of Man and the City of God) had only partial and temporary success, being eventually co-opted and recuperated by the disciplinary culture of the traditional religious hierarchies.

The history of the university, which in the beginning was, as we have seen, inextricably linked with that of the church, can in turn be regarded as the history of an institution that has repeatedly failed its liminal vocation. That the university indeed has this vocation is obvious even from its preferred geographical sites, either halfway between town and country or in a *hortus conclusus*, sheltered from daily mundane affairs. From Plato's Academy (situated in a holy olive orchard outside Athens) to the modern North American campus (often situated in a small rural town that, not infrequently, owes its origins to the land grant college itself), the university, like the monastery, has often marked and circumscribed its own ground or field. Within the confines of this field—understood in a literal sense as well, the English word "campus" being a transliteration from Latin, meaning "field"—different ground rules are supposed to obtain from those of the outside world. Common language usage that pits the academic world against the "real" world equally points to the university's liminal vocation.

There has often been an implicit or tacit pact between the university and the rest of society that has allowed the university a degree of freedom of thought and experimentation. Such freedom is seldom granted to other fields of human endeavor, say, business or politics, and it varies according to the degree of openness or closeness of a particular community or society as a whole. In return, the university was understood to function as a neutral ground where fresh ideas, sociocultural and scientific theoretical models, and even lifestyles could be safely debated and tried out before they could be introduced into society at large. This relative freedom partially explains why universities could occasionally break their pact of neutrality with the powers that be, turning into hothouses of social reform or even social revolutions.

Furthermore, the university's moral authority has also partially derived from its neutral, liminal position: the academic world, like the monastic one, has often been perceived, at least by outsiders, as a place where moral values and moral conduct are exemplary, precisely because its members are free from the constraints of the "real" world of power. Unfortunately, we have already seen how the university, no less than the monastery, has often failed to live up to its liminal vocation, merely replicating the disciplinary structures and habits that prevail outside academia.

Under the impact of globalization (understood, obviously, not as neoliberal globalizing trends, but as global awareness of the reciprocal causality of all human actions and the enhanced sense of individual responsibility resulting from such awareness), the contemporary university will hopefully be able to reform itself and regain its liminal vocation, as well as the trust that the rest of the community has placed in it. The current global circumstance presents the university with new opportunities to activate its liminal potentialities. Worldwide socioeconomic and cultural trends in the past two decades or so have increasingly revealed that disciplinary knowledge and its institutional frameworks can less and less effectively cope with the concrete social, economical, political, cultural, technological, medical, environmental, ethical, and other problems that arise in our globalized communities. There is a growing feeling inside and outside academia that fresh or revised models of knowledge and fresh or revised research and learning strategies must be developed on the part of our academic and nonacademic communities to cope with these issues. In turn, these models can no longer be monolingual and monocultural, but must take into consideration a wide diversity of cultures and systems of values and beliefs from all over the planet.

As we have seen from my discussion of contemporary cultural theories of globalization in previous chapters and from my reference, in this chapter, to some of the essays in *Universities and Globalization*, there is no consensus as to what these models should look like. But we can at least imagine what some of their fundamental principles should (or should not) be and begin to submit them to extensive intercultural dialogue and negotiation. As a preliminary step, we should continue to re-emphasize the ethical-educational function of the local–global university of the future (*pace* Newman). This function has presently taken second place to bureaucratic bottom line fiscal considerations that have virtually turned the North American and other research and "mega" universities into forms of corporate welfare and gilded intellectual sweatshops.

In this respect, it is useful to stress again, just as Newman did, the distinction between educating students *toward* the "pursuits and professions in life," to use the language of the 1862 Land Grant College Act, and training them *for* such pursuits and professions, which should be left largely in the hands of the professional schools. This does not mean that the university should expel the professional schools or the sciences from its ranks, as Newman's arguments occasionally implied—although even he "suffered" the establishment of science laboratories in his Catholic university in Ireland and, during his tenure as rector of that university, presided over the founding of the *Atlantis Magazine*, which promoted scientific research among his faculty. On the contrary, it would mean to reform the entire university in such a way that ethical-educational objectives, oriented toward global intelligence, would inform all of its components, not just the humanities, the environmental sciences, or any other individual academic field.

All of the metaphors I have so far employed to describe the university, whether borrowed from military and industrial architecture or from the world of commerce, hide or underplay the mutual causality relationship between education and ethics (understood as a particular system of values and beliefs that guide the thoughts, emotions, and actions of an individual or a community). This mutually causal relationship has lately become obscured in the free market notions of cultural capital. Or perhaps ethics itself has become such an immaterial commodity with its own fluctuating market value, as we see, for example, in the civic activities of various NGOs (nongovernmental organizations) and CSOs (civic society organizations), including proenvironmental ones, which are factored in by stock market investors in the so-called social awareness stocks. Be it as it may, most of the Western critiques of contemporary academia do agree that we need to restore the ethical-educational dimension of the university. But we should go even further and reflect on what a liminal academic model should be in the light of an emergent ethics of global intelligence, which ought eventually to become the standard for all of our educational values.

We may begin by defining the main function of the local–global university of the future. This should be neither training students for a professional career, which should be a supplementary academic objective; nor should it be a disciplinary and bureaucratic enhancement of power, nor the financial bottom line, nor a gilded sweatshop for industry and commerce, which should be no academic objectives at all. Nor should it be pursuing knowledge "for its own sake," which would ultimately lead to overweening Faustian pride. Rather, its main purpose should be to pursue cooperative learning and research in the service of human self-development. As I pointed out in the introduction, it is this kind of development that our scientific, political, academic, business, and community leaders should focus on in the new century. And, since from the standpoint of global intelligence, there are no developed or underdeveloped societies on this planet, but only developing ones, all regions of the world would be equally able and "competent" to contribute to such development. They could, for instance, be called upon to draw on their enormous cultural resources of age-old wisdom that, as we have seen throughout this study, have largely remained untapped by our current and past elites.

The very purpose and organization of education as a whole should thus be rethought, restructured, and reoriented toward global intelligence. Rethinking education in these terms will require, for example, restructuring geopolitical models based on area studies and interdisciplinary approaches in the sciences and the humanities that leave disciplinary structures largely intact. A global perspective will lead to remapping the traditional divisions of knowledge and will generally call for fresh ways of educating our younger and older generations at all levels. Indeed, it will ultimately require that learning become a lifelong process and extend well beyond formal education to all members of our local–global communities.

The university of the future can best accomplish its main objective of human (self-)development through liminal institutional models. The liminal university would share some of the features of the university as halfway house of knowledge, but without its disciplinary and commercial dimensions. Like its commercial and disciplinary counterpart, the liminal university would participate in a global network of communication and exchange. But, this network would be organized as a rhizomic structure. Its emblem would thus be the "roots of knowledge," instead of the traditional, hierarchical "tree of knowledge" (present in many esoteric, religious and nonreligious, teachings as well).

In turn, this metaphor would not relate to the origins of knowledge, which are ultimately "rootless" (because they emerge from the liminal interstices that, we recall, power-oriented mentalities conceive as nothingness), but to its mode of transmission and utilization. In this respect, the roots of knowledge should not be imagined as tentacles that reach everywhere and strangle and consume everything in their deadly embrace, somewhat in the manner of Hardt and Negri's rhizomic empire. Rather, they should be seen as free, generous, mutually nourishing and life-enhancing flows of communication and exchange. To develop the tree metaphor to its logical conclusion, society itself, which is served by the university, should be conceived as the branches of the tree. In this manner, roots, trunk, and branches engage in a symbiotic relationship of constructive amplifying feedback loops to the benefit of the entire tree, or self-organizing system.

The liminal university would not generate intellectual, or cultural, or human "capital," nor will it be part of any "service industry," even though other, nonacademic practitioners might use the knowledge it produces for such purposes. Nor would it reproduce within its framework the commercial and disciplinary relationships that obtain in the "real" world. On the contrary, it will generate and communicate, in a free and generous manner, new transdisciplinary and cross-cultural knowledge and principles of education that will be put in the service of human development.

The local–global university will work closely not only with its sister educational institutions worldwide, but also with all other nonacademic fields of endeavor, encouraging them to focus their efforts on the kinds of socioeconomic and cultural development that will in the end serve human development as a whole, throughout the planet. To this end, the university will create its own ethical practices that will be those of a liminal, honest broker, to adopt again a commercial vocabulary, rather than those of a junior or senior business "partner." It will build cross-cultural and transdisciplinary rhizomic academic networks all over the world, through which it will educate local–global citizens in the spirit of global intelligence.

Finally, the university of the future ought to mobilize all its human resources to work together toward creating a mindset conducive to alternative ways of relating to each other in our profession and in our world communities at large. If we genuinely wish to change anything in our human and natural environment, we need to begin with ourselves. To cite Oscar Wilde's witty comment on exemptionalism (*avant la lettre*), "it is so easy to convert others," but "it is so difficult to convert oneself" (Wilde 1954, p. 986).

7
Creating Local–Global Learning Environments for Human Development

In the preceding chapter, I have outlined the general principles and practices that should lie at the foundation of a local–global university. This vision informing our long-term objectives would in turn involve a number of intermediary steps. In this last chapter, as well as in the appendix, I would like to propose a program of specific reforms of our present institutions of higher learning as a modest, preliminary step toward an educational system, oriented toward global intelligence. In turn, these reforms should be part of the larger framework of creating local–global learning environments throughout the world—an issue that I shall address in the last part of this chapter. Such learning environments would equally be oriented toward global intelligence and would bring together both academic and nonacademic institutions, including local and global, governmental and nongovernmental, state and private organizations and involving persons of all ages and walks of life.

1 Preliminary Reform Program for the North American Research University

Concerning higher education, my proposals are necessarily "local," referring especially to the large, research university in North America. If these reforms were implemented, however, they would resonate positively with other academic institutions in various parts of the world, where some of these practices are already in place. Indeed, many of them were also operative in North American universities a few decades ago and have not entirely disappeared from the contemporary North American academic scene. Others may even date as far back as the first European universities, founded in the Middle Ages. One could indeed look at the medieval universities (together with the ancient monasteries) as nodes in a vast global network of knowledge that are coming again to the foreground of world history. The question, however, is not to return to these earlier practices, as if such return were possible, but to reorganize them in line with the present global circumstance.

In attempting to reorient the university toward global education, let alone global intelligence, we come up against what seem to be insurmountable obstacles, because the very academic place that has traditionally been designed to address important social and human problems seems now to compound, rather than to alleviate, such problems. As we have seen, many of our educational institutions have simply become reflections of global predicaments, instead of active leaders out of such predicaments. For instance, at most U.S. universities, current administrators, despite paying lip service to the "internationalization of the curriculum," often perceive study abroad and experiential education as expensive extras that interrupt students' normal campus activities. To make matters worse, the academic credit systems that are currently in place at most North American universities are highly protectionist. Through time- and energy-consuming bureaucratic red tape, they make it deliberately difficult for students to move across disciplines and institutions of higher learning both in the United States and overseas.

As if this situation were not distressful enough, many prestigious North American universities have increasingly become a locus of ideological strife, of warring special interest groups that blindly pursue their narrowly conceived political agendas to the detriment of the academic community at large. The situation is further aggravated by opportunistic, "neoliberal" administrators who, under the pretext of dwindling financial resources, pursue their own short-sighted policies of turning the academy into a profitable, free-market enterprise, at the expense of education, local and global. Given these egregious problems, it is imperative that new educational networks soon be put in place. The main purpose of such networks would be to bring together some of the best human and material resources of universities in order to create the kind of flexible cross-cultural and transdisciplinary curricula and organizational frameworks that are required in today's global environment.

In terms of immediate, corrective action, here I can propose, without going into much detail, only a few, obvious, measures that would help bring the North American university in line with other positive, local–global developments. The first concrete action should be for faculty members and students to reclaim the university from corporate bureaucratism. We as teachers, scholars, researchers, and advanced learners are as much to blame for the sorry state of higher education today as the corporate academic bureaucracies that have occupied our universities, with our active or passive complicity. General complaints and malaise are not going to solve the problem. We are all familiar with the political syndrome, already signaled by Socrates in Plato's *Republic*, of the reluctance of accomplished scholars and teachers to get involved in governance, allowing their supposedly less accomplished colleagues to carry out this unpleasant task. But there is no guarantee that those who do not seem to be the best academics will make the best academic administrators. So all of us should stop complaining about being misgoverned and start taking turns in governing ourselves.

We already have most of the institutions of academic governance in place, such as faculty senates, councils, committees, and so forth, which are needed to start carrying out any desired reforms. The traditional governing system of the American college, with its tripartite structure (board of nonacademic trustees, academic administration, and faculty) is not a problem in itself, even though originally it led to power struggles among these three bodies. One recalls, for instance, that the founders of the first two colleges in the United States, Harvard and William and Mary, intended for these colleges to be governed by their own members. The Harvard Charter of 1650 stipulated a corporation of president and fellows, with a board of overseers. In turn, its counterpart at William and Mary set up a corporation of president and masters, with a board of visitors. These were models of self-governance inspired by those at Oxford and Cambridge. Soon, however, the nonacademic overseers and visitors took over from the academic body the management of its affairs, on the not unreasonable assumption that universities should subordinate their narrow interests to the broader concerns of the community. To the present day, faculty members at most American universities and colleges are the employees of the nonacademic trustees.

In principle, sharing responsibility in the affairs of the university among its three governing bodies would not be harmful to its main educational objectives. On the contrary, it would ensure that town and gown work together for the benefit of all. The problem is the current imbalance, where the bureaucratic administrative branch has taken over, separating itself from the regular faculty members. In turn, the university trustees more often than not do not represent the interests of the community at large, but narrow, special interest groups, such as those of local commerce, industry, and so on. They are often appointed as trustees not because of their high moral and intellectual standing in the community, but because of their financial or political stature. One should, therefore, redress the current asymmetrical balance in favor of the faculty members, according to the well-known principle of checks and balances in government. One should finally introduce a fourth factor in this equation, that is, the equal participation of the student body in academic governance. This would, again, not mean some daring, revolutionary innovation, but only returning to the first historical model of a university, that of Bologna, where the founding corporation was that of students rather than masters.

Here are a few obvious measures that, if implemented, would go a long way toward reforming the North American institutions of higher education:

- Reform university charters to bring them in line with current democratic and global developments. Many U.S. universities have antiquated charters that are ill suited to deal with present world conditions and could, in some cases, facilitate autocratic and bureaucratic abuses.
- Reform the academic bureaucratic and legalistic system, rethinking tenure, professionalism, excellence, and other sacred cows that make the university one of the least dynamic organizations in American society. Reform the university trustee system, so that boards of trustees represent the highest ideals and broadest interests of the academic and nonacademic communities.
- For these reforms to work, remove the system of extra compensation for serving in academic administration, beyond partial release time from teaching and scholarship. Reduce top administrators' terms in office to relatively short, nonrenewable periods. *These would be the single most important measures in our universities, that will radically reduce the self-perpetuation of the academic corporate bureaucracy and its onerous red tape.* At present, top academic bureaucrats, just as their corporate counterparts, pay themselves huge salaries in comparison with regular faculty and have the greatest incentive to remain indefinitely in office, as well as to multiply and perpetuate their ranks. In order to correct this situation,

serving in academic administrations at all levels, not only on faculty committees, should become one of the routine obligations of each faculty member, no less important than teaching and research. Above all—the point bears emphatic repetition—it should not be monetarily rewarded, beyond some teaching and research release time. In order to make the process efficient, doctoral students in all fields of studies might take required courses in academic management, as well as complete an internship in an academic administrative office, so that they can become acquainted at an early stage with the organizational structure and operation of higher education and other academic institutions in North America and other parts of the world.

• Financial affairs of the university, including fund-raising, should be conducted by business professionals, hired by and directly answerable to the faculty governing body. Major budget decisions should be debated and approved by this body as well.

• Administrators at all levels, including presidents, should be elected by faculty through general elections and not by administratively appointed search committees or, in the case of top-level administrators, through the so-called head-hunting practices typical of corporate America. The criteria of selecting a university president should be his or her academic qualifications and willingness and ability to lead the academic community toward the desired educational goals, collectively agreed on. Above all, the president should do no fund-raising, which should be entrusted to professionals from the business world. A university president is not a CEO, as many presidents of North American universities like to call themselves these days. She is an academic leader well respected by her peers for her substantial contributions to teaching and scholarship and for her high moral profile.

• Allow students, through their representatives, an equal say in academic governance. But this say should not be understood in the sense of the "customer is always right"—a commercial principle that, alas, is all too often applied by our current academic bureaucrats, mindful of the bottom line. Rather, students should participate in academic affairs, not just in extracurricular activities and athletics, with an increased sense of responsibility and genuine care for their educational institutions.

• Reduce to a minimum academic regulations and red tape that are presently strangling our campuses, also with regard to the academic credit system. Students ought to be able to move freely across all fields of study and across national and international academic boundaries. They should increasingly become able to organize their own degrees across disciplinary lines according to their individual, self-development goals and academic interests.

- Do away with academic departments and create flexible centers and institutes of teaching, learning and research, according to the interests and abilities of the faculty involved and the general research topics to be addressed. Whenever possible and appropriate, such centers and institutes should engage in transdisciplinary teaching and research across the three main branches of learning, that is, physical sciences, social sciences, and the humanities, periodically changing their research focus and composition.
- Reform and further develop transdisciplinary and intercultural centers of continuing education and university extensions, to the intellectual and moral (and not primarily financial) benefit of the academic and the non-academic communities involved.

Finally, I would like to focus briefly on two practical ways in which one may begin to develop transdisciplinary and intercultural research within the current framework of the university: one is a restructured and expanded area studies, the other is the introduction of colloquia or advanced seminars on the state of knowledge in the three main branches of learning, both of which could become regular features of the advanced undergraduate and graduate curriculum.

Arjun Appadurai (1996) is illuminating on the history and the future potential of area studies. He points out that they were created at North American universities after World War II as ways of meeting the government's global strategic concerns. Area studies were designed to generate "strategically" relevant information about the politically and militarily sensitive regions of the world, including the Soviet block and the so-called Third World. After the fall of the Berlin Wall in 1989, however, area studies have gone into a tailspin.

As Appadurai puts it, the "left-wing critics of area studies, much influenced by the important work of Edward Said on orientalism, have been joined by free-marketeers and advocates of liberalization, who are impatient with what they deride as the narrowness and history fetish of area studies experts. Area studies scholars are widely criticized as obstacles to the study of everything from comparison and contemporaneity to civil society and free markets" (Appadurai 1996, p. 16). But, as he also notes, no criticism that "is so sweeping and so sudden could be entirely fair, and the odd mix of its critics suggests that area-studies scholarship might be taking the rap for a wider failure in the U.S. academy to deliver a broader and more prescient picture of the world after 1989" (ibid., p. 17).

On the other hand, some of the criticism is not entirely unjustified: area studies as a field has "probably grown too comfortable with its own maps of the world, too secure in its own expert practices, and too insensitive to transnational processes both today and in the past" (ibid.). But most important of all, one should add, this field has never done away with a disciplinary way of thinking: instead of being transdisciplinary, it has at best remained interdisciplinary. As such, it continues to operate very much in terms of disciplinary turfs, expertise, national prestige, strategically relevant research topics that generate federal funding, and so on. This is also the case for the mushrooming centers of global studies that have supposedly replaced area studies at a number of North American universities, but for the most part perpetuate exactly the same disciplinary assumptions and practices.

Even so, both area studies and global studies, with their emphasis on the need for advanced study of foreign languages and cultures, alternative worldviews, and macroperspectives on sociocultural change in various regions of the world can be regarded as a promising bridge toward transdisciplinarity within the fairly parochial intellectual environment of the mainstream North American university. They should therefore not be dispensed with, but on the contrary, used as an existing valuable resource for building genuinely transdisciplinary and intercultural research programs, informed by principles and practices conducive to global intelligence.

Another such transdisciplinary bridge, this time between the three main branches of learning, might be built with the help of a series of colloquia or advanced seminars on the state of knowledge in natural sciences, social sciences, the humanities and the arts throughout the world. These colloquia or seminars would be part of required coursework for all students, regardless of their major field of study and would carry full academic credit toward the terminal degree at both the undergraduate and graduate level. They would examine the basic methodological questions specific to the three main disciplinary cultures, would feature their major recent advances, would explore differences and similarities, as well as establish new transdisciplinary connections among them.[1]

The seminars/colloquia would include up to twenty-four student participants evenly distributed among different fields and, whenever possible, different cultural backgrounds; three coinstructors chosen from the regular faculty specializing in one of the three main branches of learning; and three or six distinguished guest speakers, also representing one of these branches and, whenever possible, more than one culture. The seminars would also be divided into three rounds of an appropriate number of sessions, plus an introductory and a concluding session.

The introductory session would explain the objectives and methodology of the exercise and would distribute the various tasks among the participants, dividing them into three teams, each focusing on one of the main branches of knowledge. The first round of sessions would in turn be divided into three parts, one concentrating on the physical sciences, the second one on the social sciences, and the third one on the humanities. The second round would contain discussions of three collaborative reports presented by each of the teams on the state of the three branches of knowledge. The third round of sessions would concentrate on generating a comprehensive report on all of these fields with concrete proposals designed to cross-fertilize and enrich them.

Of course, there are already a handful of universities in the United States and elsewhere that have begun to introduce innovative, cross-disciplinary, and cross-cultural curricula in response to today's imperatives of global education. One such university is the newly founded U.C. Merced, the tenth campus of the University of California system. This university proposes to encourage and support extensive links among the various disciplines, whether in the humanities or the sciences, keeping the current barriers between them as low as possible. Students, not just faculty, will equally be expected to get involved in research conducted at the university, beginning as early as their undergraduate years. In turn, research is organized around crosscutting, multidisciplinary institutes, including the Sierra Nevada Research Institute, with a focus on environmental studies and policy, and the World Cultures Institute, with a focus on issues of global population movements and their historical and cultural consequences. All of these institutes are expected, wherever possible, to coordinate their research, addressing community needs and sustainability challenges.

A crucial aspect of U. C. Merced and the planned community developments around it is an environmental architectural design that uses the latest advances in clean technology to create an ecologically sound educational and living ambience. This is particularly significant in view of the fact that many of the current "mega" universities are also mega-wasteful consumers of energy and other natural and human resources. Ironically, many of them show little concern (beyond lip service) for the impact of their research and other activities on the local environment, despite their copious course offerings in environmental studies.

Other examples of new institutions of higher learning that seem globally oriented and animated by an innovative learning spirit include the International University of Bremen in Germany, Sabanci University in Istanbul, Central European University in Budapest (with research networks in several East European countries), Venice International University in Italy (with research networks in Spain, France, and the United States) and EARTH University in Costa Rica (with an impressive international faculty), to mention just a few. There are also transnational educational efforts on the part of the European Community, such as the Socrates program. These efforts should be coordinated not only with transatlantic educational efforts (as in the ECUSA program), but with worldwide educational initiatives, especially in terms of academic credit transfer and free movement of students across all national and disciplinary boundaries, not just within the Western world.

In the end, however, the success of these educational experiments will depend on the ability of the administrators and the teachers involved to go beyond the current bureaucratic and disciplinary habits that are very difficult to break, whether in academia or outside it. Judging from the failure of earlier experiments such as the United Nations University, which could have set the paradigm for global education, but which has turned into an uninspiring, bureaucratic academic network, one could only hope that at least their alumni, if not their founders, will live up to the ideals embodied in these worthwhile learning experiments. And, of course, there is no good reason why the United Nations University system itself could not undergo extensive reforms in the spirit of global intelligence. In fact, both the UNU system and UNESCO would be in an excellent position to spearhead precisely these kinds of reforms.

2 Local–Global Learning Environments for Sustainable Human Development

As I have already mentioned, substantive reforms of the North American and other universities can be accomplished only by concerted action of reform-minded academics from inside academia and civic groups from outside it. To this end, it would be crucial to embark on an extensive dialogue at all levels of North American society on the future role of the university in a local–global environment and on the nature of the reforms to be implemented. One could, for example, create CSOs (civil society organizations) that would monitor and work with universities and other educational institutions to reform North American higher education. One could also establish, say, an "education watch" to bring some of the more egregious abuses of corporate academia to the attention of the press and the general public, as well as to debate and recommend extensive educational reforms.

One should coordinate such networks with those existing at other educational levels, both inside and outside the academia. For example, there are many professional organizations, private and public, on-line and off-line, that are concerned with developing new learning strategies in a global environment. Academic institutions should seek active cooperation with such organizations, so that mutually beneficial synergies can develop. All of these learning networks could, in turn, enter in resonance with each other. Most important, they could also enter in resonance with the worldwide ecological movement, especially with its deep, nonviolent strands, of which they could eventually become an integral part, and vice versa.

Above all, as I have already suggested, we need to create local–global learning environments, understood as liminal spaces for intercultural research, dialogue, and cooperation, based on an emergent ethics of global intelligence. In the remaining space of this chapter, I would like to sum up, and in turn submit for intercultural research and dialogue, the preliminary steps that I believe we could take, inside and outside the academic world, toward accomplishing this goal.

(A) Local–Global Systems of Values and Cultural Blueprints for Human Development, Oriented toward Global Intelligence

We need to create the proper local–global environments within which the various local systems of values and beliefs can come into contact and engage in a genuine dialogue. This intercultural dialogue will not be devoid of conflicts. But it is precisely the willingness to negotiate such conflicts in peaceful, amicable, and mutually advantageous ways that might ultimately lead to the creation of local–global systems of values, grounded in a common human ethos, that all of our world communities can freely agree on. This common human ethos may, for example, include such widely accepted values as care for the human and natural environment, respect for and delight in cultural differences, responsive understanding and interaction, peaceful and mutually beneficial cooperation, need for spiritual transcendence, generosity, love, compassion, and so forth.

A closely related goal would be to create local cultural blueprints to support and sustain the local–global systems of values. These cultural scripts would obviously not be imposed from the global "outside," if indeed such concepts as "outside" and "inside" were still operative in our age. Instead, they would spring from and resonate with the most cherished aspirations of a local community. They could start from, but then develop and reform, or even partially replace, the current local, nationalist, ethnic, and racial core of values and beliefs. In this regard, it is instructive to recall, as Benedict Anderson (1991), Mike Featherstone (1995), and others do, the ways in which ethnic and national communities were invented in eighteenth- and nineteenth-century Europe. There were deliberate and sustained attempts by cultural practitioners such as literary and art critics, playwrights, poets, novelists, men and women of letters, composers, painters, actors, journalists, magazine reviewers, translators, grammarians, schoolteachers, and educators at all levels to create national scripts based on a common repository of myths, heroes, events, landscapes, memories, customs, and other popular cultural resources.

Such attempts were in turn supported and amplified by the availability of a print culture that could bring people together over time and space. As Featherstone notes, the creation of a nation thus depended on "the development of the book, the novel, and the newspaper alongside a literate reading public capable of using these resources throughout the territorial area and thus able to imagine themselves as a community" (Featherstone 1995, p. 53f). One may add to the list such cultural institutions as the national theater and opera, literary and artistic circles and associations, and so forth, that promoted the national agenda. Featherstone mentions also the development of the film industry as an important medium of cultural propagation. There is no reason why our contemporary media cannot contribute to a global educational program that will resonate with various local traditions. Indeed, information technology and global communication can play a decisive role in promoting the local–global core of values.

(B) Making Judicious and Responsible Use of Our Information and Communication Technologies

We now obviously possess advanced means of communication that continue to develop at breathtaking speeds. But, the important questions remain the old ones: What is the content of our communications? What uses are we putting our new information and communication technology to? We have seen that post-Marxist cultural critics rightly deplore the current role of the new media and information technology in promoting Western-style consumerist values all over the world (although we may wonder, with postmodernist and other scholars, about just how successful or effective they are in doing so). Yet criticism and self-criticism of the new media, although perhaps necessary as a first step, will ultimately remain empty gestures, unless we find ways in which we can put communication and information technology to culturally productive uses. Finally, the point is not "resisting" the World Wide Web—it would be like resisting the weather—but creating communication networks to educate those involved in the new information technology, beginning with our local–global media elites.

One may launch educational initiatives on the Internet that could mediate between the various cultures and mindsets that participate in this vast network, helping its members to become responsible world citizens. Rather than resisting distance learning and other technological aids to instruction because they are supposed to be dehumanizing, let us humanize them by changing the very nature of what we communicate through them. In turn, let us create new, intercultural-friendly types of software that will respect and nourish global cultural diversity and will replace our current standardized types, based on monological and monocultural principles.

(C) Making Judicious and Responsible Use of All Other Technosciences, as Well as of Science as a Whole, to Create Sustainable Environments for Human Development

As deep ecologists and other scholars, scientists, and practitioners point out, our current mainstream commercial, industrial, and ecological practices remain unsustainable, both in the West and in other parts of the world. Paul Hawken (1993), Hazel Henderson (1999), and Ervin Laszlo (2003) among others, list a number of actions that our world communities can initiate throughout the planet to alleviate and reverse some of the disastrous ecological practices of the past. The most urgent ones would be to:

- Reduce emissions of CO_2 and other greenhouse gases into the atmosphere
- Reforest denuded lands and prevent erosion of cultivable lands
- Significantly reduce and clean up pollution
- Develop alternative sources of energy and lightwave technologies
- Inhibit conspicuous consumption and introduce worldwide recycling practices
- Facilitate and stimulate environment-friendly business ventures or "natural capitalism," while discouraging wasteful and ecologically destructive business practices
- Provide sustainable living, learning, and working conditions for women and children, as well as the underprivileged in general
- Encourage a reverse flow of people from the cities back to the countryside
- Retrain the unemployed and the underemployed
- Do away with weapons of mass destruction and all other dangerous technologies
- Reallocate resources in favor of education and healthcare
- Encourage and support smaller businesses, smaller educational and other private and public institutions, smaller bureaucracies, etc., according to the principle of "small is beautiful" (Schumacher 1975)

At present, our mainstream technosciences such as biotechnology are part of the global problem, rather than the solution. We need therefore to develop an ecology of science, as well as an ecology of learning as a whole, grounded in the spirit of global intelligence (rather than in exemptionalism), as urgently as we need to address the problems listed by Hawken, Laszlo, Henderson, and others. Only local–global learning environments supported and amplified by sustainable forms of technoscience can help our world communities become aware of the urgency of such problems, as well as address them in an effective and mutually beneficial manner.

Information technology and other new technologies that are based on the principle of mutual causality or on a nonlinear mode of being and acting in the world will probably fare better in future global circumstances than the ones based on linear models. Ultimately, however, information technology, or any other technology, will be only as "good" (or as "bad") as the intentions and objectives of the humans who use it. For that reason, contemporary technoscience, together with all other science and technology, should primarily evaluate itself—and be in turn evaluated and promoted—in terms of its potential contributions to overall human development through global intercultural learning and research, rather than in terms of their short-term contributions to material well-being and/or profit of a relatively small fraction of humanity. In other words, it should evaluate itself, and be evaluated by society at large, according to the emergent ethics of global intelligence.

(D) Remapping Knowledge and Reorienting Our Educational Systems toward a Global Reference Frame

Our current educational systems are rooted in the nineteenth-century transition from agricultural- to industrial-based economies and the creation of the modern nation-state. Therefore, they have largely been structured to prepare our youth for citizenship, employment, and a moral and productive life within the nation-state, focusing mostly on the national economy, security, and welfare. But, our growing awareness of the planet as an interdependent web of life moves us increasingly toward an entirely different world, in which old disciplinary boundaries, whether national or cognitive, have largely outlived their purposes.

It is small wonder then, that most of our current centers of education and research can hardly keep up with such transdisciplinary and transnational developments. Education and the very purpose and organization of our academic institutions must now be rethought and restructured within a global reference frame. Rethinking education in local–global terms will require remapping the old disciplinary divisions and will generally call for new ways of educating the local–global citizens of tomorrow. Indeed, it will ultimately require that learning become a lifelong process and include all members of our world communities, young or old. Under the impact of lifelong learning, these communities will ideally become genuine laboratories (and playgrounds) of cooperative, intercultural discovery and creativity.

(E) Creating Worldwide Intercultural Educational and Research Pilot Programs to Promote Local–Global Learning Environments

We need to design advanced intercultural and transdisciplinary academic programs that employ educational and training strategies appropriate for a global age. Such programs should start from the recognition that many universities have valuable human resources, academic and nonacademic networks, and in some cases excellent infrastructures, such as modern laboratories, libraries, electronic equipment, distance learning facilities, and so on. Many are, moreover, located in cosmopolitan centers of culture, attracting some of the best theoretical and practical minds from around the world.

Because universities could and should play a crucial role in global education, it is not so much a question of creating entirely new institutions than one of redesigning and reorienting the current ones toward global intelligence. For that purpose we need experimental pilot programs that would bring together some of the best human and material resources of universities and other academic and nonacademic organizations from several continents in order to remap traditional knowledge, generate new kinds of knowledge from a global perspective, and design the kinds of flexible cross-cultural and transdisciplinary curricula and practica that are required in today's global environment. In the appendix to the present book, I propose one such pilot project: an advanced academic program in a new transdisciplinary and cross-cultural field of studies that I have called Intercultural Knowledge Management. This program will provide a select number of outstanding students, teachers and practitioners from various parts of the world with an appropriate local–global environment for teaching, learning, and research, based on the emergent principles of global intelligence. If such projects can be implemented and will prove successful, they will undoubtedly inspire many other such global educational initiatives and will ultimately stimulate and support extensive reforms in world education.

(F) Creating Local–Global Elites in the Spirit of Global Intelligence

Many cultural theorists and practitioners believe that at present our world communities undergo a crisis of leadership. According to them, many leaders no longer serve their communities but only themselves and their narrow, inner circles. Globally and locally speaking, however, the situation is again much more complicated, in the sense that there are many dedicated and well-meaning leaders from around the world who genuinely attempt to represent, and act according to, the best interests of their communities. They are often at a loss, however, in determining what this best interest might be, confronted as they are with both internal and external political and economic pressures, power struggles, conflicting advice from policy experts, and above all with the perplexities of a complex, interdependent, and largely unpredictable global flow of material goods, information, people, images, and ideas.

Therefore, we need to create pilot global learning programs such as the Intercultural Knowledge Management project also because our future leaders urgently need to get an education appropriate for the increasingly complex demands that a globalized world will make on them. In the end, however, we should remember that our leaders, just like our technologies, are only as "good" or as "bad" as we are, in other words, that communities and their leaders are also engaged in a relationship of mutual causality, with productive or destructive amplifying feedback loops.

(G) Connecting Our Worldwide Educational Networks with Other Global and Local Networks

These are nongovernmental or civil society organizations, public and private institutions, community service organizations, and so forth. Hazel Henderson (1999) lists a large number of such organizations worldwide that would be compatible with and could help develop the principles and practices of global intelligence. Of course, many more such organizations could be added to the list. In principle, however, educational networks oriented toward global intelligence should work not only with like-minded academic and nonacademic organizations, but also with those that seem incompatible with them. A meaningful and productive dialogue, based on peaceful, responsive understanding, should engage members of all social and cultural groups, irrespective of their ideology or political beliefs. One of the most counterproductive human attitudes is sectarianism of any kind, be it ideological, political, religious, or cultural, which is yet another form of exemptionalism.

A key to reforming world education is to bridge the current gap between the academic and nonacademic worlds by joining the practical experience and proven success of leaders from the private and the public sectors to the strategic reflection and openness to creative ideas that guide the work of prominent scholars, artists, and thinkers from around the globe. Ideally, the university should become a part of the whole community, and the whole community should become part of the university, so that young and old can learn together and from each other throughout their lives.

(H) Global Learning Forums

One way of connecting our global learning networks would be through regular international forums on education, convened in various parts of the world. They would debate the educational reforms needed in various countries and regions and would put forward proposals of such reforms, to be implemented by local communities throughout the globe, according to their specific cultural and educational needs. It is essential that these forums include participants from all walks of life, representing large cross sections of the world population. Education is just too important for society as a whole to be left solely in the hands of education "experts," whether academic or governmental, just as science is too important to be left in the hands of scientific "experts." Both education and science should therefore become the object of community-wide and worldwide, intercultural dialogue, research, and cooperation.

In conclusion, this book has attempted to outline a number of principles and practices that could facilitate the emergence of global intelligence. These principles and practices are not new, but are available in most, if not all of our traditions of wisdom. They have often been neglected, distorted, or pushed into the background by a mentality of power that has prevailed on our planet for a few thousand years. Although to us, individual human beings, this period seems very long and may lead us to believe that power has always been and will always be the organizing principle of humankind, it is nevertheless only "a tick in geological time," as E. O. Wilson (1998) puts it. Seen from the "perspective of the universe" (Sidgwick 1874), humanity is only at the very beginning of its development, which should give us cause for hope, rather than despair. We seem to be approaching what evolutionists call an evolutionary bifurcation, and I have implied that what we call "globalization" could be regarded as one of its preliminary manifestations.

Even though at present globalization has assumed at least as many negative as positive forms, it remains fraught with promise, if we understand by it the emergence of a collective awareness that our globe is an interdependent web of life, in which we human beings have a decisive role to play. As many ancient myths of genesis tell us, including those in the Old Testament and the *Tao Te Ching*, we are the stewards of this planet. Unfortunately we have so far interpreted this message mostly as a license for exemptionalism, rather than as an immense responsibility toward each other and all other life on earth. It is largely this exemptionalism that I have called a "mentality of power," in the wake of countless teachers of humanity throughout our collective world history.

To fulfill the promise inherent in our evolutionary capabilities, particularly in the present global circumstance, we need to work individually and collectively toward changing our power-oriented mentalities, that is, the exemptionalist ways in which we relate to each other and to our environments. This change, I have suggested, would imply refashioning ourselves, our communities and, generally, life on earth according to irenic organizing principles. Such a radical shift constitutes a century-long objective that can be accomplished only through a sustained, unprecedented, worldwide collective effort on the part of all of our global communities.

Some readers will undoubtedly regard my present arguments as an exercise in utopian thinking. If so, I would like to remind them of Oscar Wilde's remarks about his own utopian speculations in "The Soul of Man under Socialism":

> It will, of course, be said that such a scheme as it is set forth here is quite impractical and goes against human nature. This is perfectly true. It is unpractical and it goes against human nature. This is why it is worth carrying out, and that is why one proposes it. For what is a practical scheme? A practical scheme is either a scheme that is already in existence, or a scheme that can be carried out under existing conditions. But it is exactly the existing conditions that one objects to; and any scheme that could accept these conditions is wrong and foolish. The conditions will be done away with, and human nature will change. The only thing that one really knows about human nature is that it changes. (Wilde 1954, p. 1039)

Of course, other proposals for further human development would be equally appropriate and desirable. I hope that my present appeal to the world community of scholars and practitioners to engage in intercultural research, dialogue, and negotiation on these and other local–global issues will not go unanswered and that all of us in academia will soon begin to work together toward these or similar reforms. Above all, I hope we will work toward restoring the liminal vocation of the university and of all our branches of learning. For that arduous task, we need temporarily to suspend our professional skepticism and critical "second nature," perhaps even our hardened sense of reality, which all too often is only a hardened sense of defeat. Then we shall be ready not only to dream about, but also to actualize what humanity—and human knowledge—can truly be and accomplish.

Appendix
The Intercultural Knowledge Management Program: A Pilot Project in Global Learning and Leadership

1 Preliminary Remarks

In this appendix, I shall put to work some of the principles and strategies for human self-development that I have outlined elsewhere in this book and shall propose the creation of an advanced academic program in a new field of study and practice, called Intercultural Knowledge Management (IKM). I shall also present a model of curricular and institutional organization that will serve the purposes and objectives of this program. I shall deliberately use the concise format and technical language of an academic proposal. Thereby, I wish to convey the concrete and practical nature of this project that could be implemented, with some modifications, even under the present state of world education. I apologize for repeating some of the points I made in previous chapters, but I wish to make this proposal as complete and freestanding as possible.

The IKM program is substantially different from the doctoral programs currently offered at our universities, insofar as it goes well beyond the scope of one or two disciplines at one or two universities. It is based on a worldwide network of academic and nonacademic institutions, in which learners, scholars, researchers and practitioners will work together toward generating local–global, transdisciplinary and intercultural knowledge. Since the creation of the IKM doctoral program will involve not only working out institutional arrangements among host universities that will form an IKM academic consortium, but also building local–global learning environments throughout the world, I shall refer to this entire development process as the IKM pilot project.

I have chosen the word "management" in the title of the program in a deliberate manner. I am fully aware that it will evoke the utilitarian and bureaucratic, managerial style that many of us deplore in our universities and professional business schools, indeed, in the neoliberal attitudes of many members of today's world elites in general. I believe, however, that we should reclaim this term for a different way of organizing and conducting human affairs, in line with an ecology of human development, oriented toward global intelligence. Creating and managing local–global forms of knowledge that will further worldwide, sustainable human development should ultimately become the goal of all of our leaders, whether in business, education, politics, or any other form of human endeavor.

I fervently hope that, in a not-too-distant future, a group of outstanding universities from various parts of the world will have the creative vision, as well as the political and administrative will, to pool their resources together in order to help launch this kind of project. But, should they decide to become involved, they will be truly successful only if they preserve its unity of vision and cooperative spirit, in other words, only if they themselves will strive after global intelligence, which is the long-term educational objective of the program.

Finally, I would like to thank numerous friends and colleagues who have at various stages participated in the development of the IKM concept and/or have offered helpful critiques and suggestions for the present proposal. I am particularly grateful to Jacques de Pablo Lacoste, Mikhail Epstein, Wlad Godzich, Wolfgang Iser, Riel Miller, Brian Rosborough, and Hardy Schloer for their kind and generous assistance with this and other global learning projects.

2 General Considerations

The rapid pace of change in the past few decades presents humankind with exciting opportunities. But rapid change can also be socially and culturally destabilizing. Problems have become highly complex, nonlinear, cross-disciplinary, and transnational in nature, requiring the best innovative solutions on the part of our communities in order to achieve sustainable and peaceful patterns of human development. Therefore, higher education and the very purpose and organization of our academic institutions must also be rethought and restructured in order to meet today's global challenges and opportunities. Rethinking education within a global reference frame will, for example, require restructuring geopolitical models based on a Cold War model of area studies and interdisciplinary approaches that leave the traditional disciplines largely intact. A global perspective will lead to remapping the old disciplinary divisions and will generally call for new ways of educating the elites of tomorrow.

The doctoral program in Intercultural Knowledge Management is the first intercultural and transdisciplinary academic program in the world that is designed with a global perspective in mind and that employs educational and training strategies appropriate for a global reference frame. The entire IKM pilot project, of which the IKM doctoral program is a part, starts from the recognition that many universities have valuable human resources, academic and nonacademic networks, and in some cases excellent infrastructures, such as modern laboratories, libraries, electronic equipment, distance learning facilities, and so forth. Many are, moreover, located in cosmopolitan centers of culture, attracting some of the best theoretical and practical minds from around the world.

Thus, universities could and should play a crucial role in global learning and research. For this very reason, the IKM pilot project does not envision the creation of yet another international university in competition with other such institutions. On the contrary, it proposes to bring together some of the best human and material resources of universities and other academic and nonacademic organizations from several continents in order to remap traditional knowledge, generate new kinds of knowledge from a global perspective, and design the kinds of flexible, intercultural and transdisciplinary, curriculum and practicum that are required in today's global circumstance.

The IKM pilot project is based on the premise that our best hope for a productive and peaceful future lies in the creative resourcefulness, openmindedness, and moral integrity of globally oriented leaders and practitioners of all ages and in all cultures. It sees itself as a global learning experiment that will provide a select number of outstanding students, teachers, and practitioners from various parts of the world with an appropriate intercultural environment for teaching, learning, and research. Should the IKM project prove successful, it would undoubtedly inspire many other such global learning initiatives and would ultimately stimulate and support extensive reforms in world education.

3 Intercultural Knowledge Management: A New Field of Study and Practice

As a field of study and practice, Intercultural Knowledge Management (IKM) is by its very nature cross-cultural and transdisciplinary. It involves not only remapping traditional knowledge, as it is acquired, accumulated and transmitted by various academic disciplines, be they scientific or humanistic, but also generating new kinds of knowledge within a global, intercultural reference frame.

IKM employs a local–global cognitive approach. This approach implies the recognition that there are many levels or reference frames of reality with their own logic and operating principles. In turn, these frames are interactively connected, affecting each other according to what general systems theorists call "mutual causality" and early Buddhist practitioners call "dependent origination." As we move from disciplinary to interdisciplinary and then to transdisciplinary reference frames, as well as from monocultural to intercultural to transcultural or global ones, new levels of reality emerge, as well as new kinds of knowledge. At its broadest theoretical level, IKM explores such questions as: What are the conditions of the possibility of the emergence of intercultural and transdisciplinary knowledge? How does such knowledge differ from, but also involve, cultural or disciplinary knowledge? How can it be communicated or taught? What uses can this kind of knowledge be put to and whom does it serve? What organizational and institutional forms might it take?

Because IKM is a new field of study and practice that is only at the beginning of its development, my concern here is less with exploring and contextualizing such complex theoretical questions, which will be the proper task of the IKM researchers and learners themselves. Rather, my concern is with sketching the general conditions and the institutional framework that would allow this field to emerge. One expedient way of accomplishing this task and at the same time moving the field forward is to lay out the concrete IKM doctoral program, with its mission, objectives, targeted audience, basic curriculum, research programs, methodology, and logistics.

I should nevertheless point out that IKM employs a model of knowledge that is entirely different from the current disciplinary ones: according to disciplinary thinking, one must first constitute the discipline, that is, an organized body of knowledge, before one can teach it, for instance, through a doctoral program. Such doctoral programs serve the purpose of both codifying the study and practice of a field of knowledge through disciplinary standards and requirements and of transmitting this code to a body of students who will in turn contribute to consolidating and expanding the disciplinary knowledge and practice that have been passed down to them.

In other words, in disciplinary models, knowledge is first acquired (learned) and then transmitted (taught). In the model of knowledge as emergence that I propose, learning and teaching are codependent and simultaneous processes, so that the field of intercultural knowledge management co-arises with the academic program that codifies, or rather continuously recodifies, its practice. Consequently, in the IKM doctoral program, teaching becomes learning and learning becomes teaching, as new knowledge continuously emerges and is continuously codified and recodified.

The model of knowledge as emergence obviously requires institutional frameworks that are different from the ones that are currently in place in our universities. Therefore, it might be useful, before I actually describe the content of the IKM doctoral program in any detail, to outline and discuss at least one possible model of the kind of worldwide institutional framework that needs to be put in place for the IKM pilot project to be successfully implemented.

4 Institutional Framework for the IKM Project

Because the IKM pilot project involves a global network of learners, teachers, and practitioners that will cooperate toward creating global learning environments throughout the world, it will require a complex, but supple and flexible institutional framework. Undoubtedly, there are several suitable institutional models one can think of, but here I would like to propose a relatively simple one, based on the continuous interaction and cooperation between two intercultural organizations: an IKM academic consortium and an Institute for Intercultural Learning and Research, which will be administratively independent from, but will work very closely with, the IKM academic consortium. This dual, interactive organizational format is essential, because it will ensure that the IKM programs operate both inside and outside the university environment, with the purpose of bringing the academic and the nonacademic worlds together within a larger reference frame, that of global learning and knowledge production.

The IKM academic consortium (the Consortium, for short) will be formed by a number of prominent universities and research institutes from various countries. These can be referred to as "host institutions." Initially, the Consortium should bring together at least eight host institutions from as many regions or countries, representing the largest cultures of the world: China, Europe, India, Latin America, Middle East, North America, Russia, and sub-Saharan Africa. Once the IKM project is well advanced, the Consortium could expand to include many other academic and research institutions from all over the world. Or, better yet, other consortia could be formed, based on the same or kindred, local–global learning principles and objectives, which could then join and/or cooperate with the initial network.

The role of the Consortium will be to host the IKM doctoral program and its participants at the member academic and research institutions for appropriate periods and to assist the Institute for Intercultural Learning and Research (the Institute, for short) in organizing the IKM curriculum, research projects, and other activities. The Consortium will have a president and a board of directors, elected from among the most prominent administrators at the member institutions. The president and the members of the board will serve three-year terms each, to ensure that all participating institutions will be adequately represented, based on a rotation principle.

In turn, the primary role of the Institute will be to organize and operate the IKM doctoral program (and eventually other research and educational programs) for the Consortium, as well as to grant the doctoral degree in intercultural knowledge management on its behalf. Of course, each of the members of the Consortium may grant its own doctoral degree to some or all of the IKM doctoral candidates, if it so chooses. For example, Chinese or Russian candidates may receive the doctoral degree from the Institute, but also from one of their own national institutions, participating in the Consortium. This dual degree strategy will address the complicated administrative problem of degree accreditation and equivalency that is endemic to all national education systems.

The Institute will be managed by a cross-cultural board of trustees, serving for three-year terms. The trustees, in consultation with the board of directors of the Consortium, will appoint an academic director and an executive director for the Institute. In consultation with the directors of the Institute and the Consortium partners, the trustees will also appoint a diverse, cross-disciplinary, and intercultural core faculty for the IKM doctoral program. This core faculty will be selected primarily from among the most prominent faculty members of the host institutions, but may include internationally prominent scholars and practitioners from other academic and nonacademic institutions as well.

The Institute will also negotiate release time and other remunerated contractual arrangements with the home departments and institutions of the IKM core faculty members. All faculty members will serve a four-year, renewable term on the IKM core faculty, after which they will return to their home institutions. They will nevertheless remain permanent fellows of the Institute and will be invited to participate in its various activities for the rest of their careers.

The academic director of the Institute must be an outstanding teacher and researcher who may be recruited from among the most prominent members of the IKM core faculty or be hired from outside. He or she will be in charge of the academic and research development of the IKM doctoral program. The executive director must in turn be a prominent scholar and should also teach in the program. She or he will be primarily in charge of admissions, logistics, finances, and public relations of the IKM program. Both directors will serve five-year terms, after which new directors will be appointed either from different host institutions or from outside the Consortium. The directors will work very closely together to ensure the smooth functioning of the entire operation.

In developing the IKM doctoral program, the directors will be assisted by an honorary board, advisory committee, ad hoc admissions committee, and ad hoc curriculum and research (C & R) committee. The honorary board will be composed of distinguished international personalities, including prominent figures from the private and the public spheres, such as global financiers, businessmen, former statesmen and diplomats, internationally known scientists, humanists, artists, and so forth. This board will help with the public relations and fund-raising aspects of the IKM project, as well as with identifying the best candidates for the IKM doctoral program throughout the world.

The advisory committee will be composed of prominent administrators, scholars, and researchers at the various host institutions and the Institute. This committee will guide and support the directors through the planning stages of the IKM doctoral program.

In turn, the admissions committee will be formed of prominent members of the IKM core faculty and student representatives. It will work closely with the local nominating committees from various countries, the honorary board, and the C & R committee of the program to identify and select the best IKM doctoral candidates.

The C & R committee will also be drawn from the IKM core faculty and student body, and will be headed by the academic director. The regular members of the C & R committee will serve two-year renewable terms. Their function will be to help develop new IKM courses, help select students for the program and help organize their course of study. Each prospective student will be reviewed by at least three members of the C & R committee, including the academic or the executive director. The C & R committee will approve courses of study for individual students, as well as an appropriate advisory committee for each student, in consultation with the student's major advisor/professor.

The Institute will operate through local centers for intercultural learning and research (local centers, for short). The local centers are cooperative endeavors between the Institute, the Consortium, and other neighboring local institutions, such as universities, colleges, research centers, private, public, and nongovernmental organizations, and so on. Their role is to conduct educational, research, and public activities in line with the programs, aims, and goals of the IKM pilot project.

The local centers recruit applicants for the IKM doctoral program and organize and oversee the academic and nonacademic activities of IKM students during their residence in a particular region. They also organize worldwide, transdisciplinary and intercultural seminars, workshops, and teleconferences through interactive, communication technology, with distinguished participants from the host institutions and invited practitioners. Additionally, they organize the introductory summer sessions in Intercultural Knowledge Management that are part of the process of admission to the IKM doctoral program.

The local centers may be located on the campus of the IKM host institutions. They will, however, be administratively independent from those institutions, while closely cooperating with them. Each local center will have a director (who will also be a member of the IKM core faculty), an administrative assistant, and a lab technician. Its infrastructure will include the administrative offices, meeting rooms, and a teleconference and video production (TVP) lab for the IKM doctoral program in that part of the world. The local centers will also house, on a rotational basis, the offices of the academic and executive directors of the Institute and their administrative assistants.

5 Mission and Objectives of the IKM Doctoral Program

The IKM doctoral program will educate a highly select group of civically oriented global practitioners and entrepreneurs. It will train them how to produce, as well as how to recognize and manage, new forms of knowledge and competencies in a global intercultural environment. The program will heighten the awareness of students that intercultural project management and problem solving involve an integrative, transdisciplinary approach that takes into account the political, social, economic, and cultural conditions, as well as the systems of values and beliefs, of local communities from around the world. It will also develop the students' curiosity about, exposure to, and understanding of each other's cultures and will teach them how to live, communicate, and work with each other in culturally diverse environments.

The IKM doctoral program will educate local–global elites who: (1) possess a thorough understanding of and a strong sense of responsibility for the "local"; (2) care for the natural and human environment, and respect and encourage cultural and biological diversity; (3) are deeply committed to seeking peacefully negotiated solutions to conflicts and have the ability to bring about such negotiated solutions; (4) know how to operate in a culturally diverse environment and across disciplines and professions; (5) develop more than one career track in a lifetime, pursuing lifelong learning; (6) comfortably serve in both the public and the private sectors and know how to generate new employment and ways of wealth making, based on wise management of the planet's human and natural resources; (7) and generally engage in lifelong creative and meaningful activity that is both service oriented and personally fulfilling.

Some of the intercultural skills and talents that IKM graduates will develop amount to what the U.S. Academic Council on Education (ACE), in a position paper entitled "Beyond September 11: A Comprehensive National Policy on International Education" calls *global competence* and *global expertise*. The paper calls for the creation of U.S. "global experts in foreign languages, cultures and political, economic and social systems throughout the world" (ACE 2002, p. 2). Global competence and expertise are certainly very important talents and skills to be developed in our national citizenry and workforce. But, for the ACE the operative word remains "national." While it deals with global issues, the ACE adopts a national and international, rather than a global perspective on these issues. A local–global approach will take into consideration not only the perceived national or "local" interests of the United States or any other country or region. Of course, those local interests are extremely important, and genuine global practitioners will neglect them only at their peril. But such practitioners will also look beyond what might turn out to be short-term and limited national interest to long-range interests, serving the entire global community. From this global perspective, the concept of national interest itself may gain a new dimension and be redefined, in a larger reference frame, as that which ultimately is in the best interest of and benefits all nations and cultures.

Therefore, in addition to global competence and global expertise, the graduates of the IKM doctoral program will strive after *global intelligence*, or the ability to understand, respond to, and work toward, what is in the best interest of and will benefit all human beings and all other forms of life on the planet. This kind of responsive understanding and action can only emerge from continuing intercultural dialogue and negotiation, in other words, it is interactive, and no single national or supranational instance or authority can predetermine its outcome. Thus, global intelligence or intercultural responsive understanding and action are what contemporary nonlinear science calls emergent phenomena and involve a lifelong learning process.

The first and main objective of the IKM doctoral program, then, is to develop in its graduates a sense of mission and commitment to local–global human values and ideals, as well as a willingness and ability to translate them into practice throughout the world. The objective of developing global intelligence or intercultural responsive understanding and action (in addition to global competence and expertise) is what distinguishes the IKM doctoral program from many professional programs with an international and intercultural focus, such as can be found at some of the top international business schools in various parts of the world. For the most part, the main objective of such international programs is precisely to develop (utilitarian) global competence and expertise that their students will, in turn, place in the service of individual private or public organizations, irrespective of the mission and goals of those organizations. Again, these are very important skills, so the ACE is right in pointing out the grievous shortage of such global experts in the United States and, one may add, in many other parts of the world. But, for a genuine local–global practitioner these skills and talents cannot be separated from the mission, goals, and ethics of global intelligence, from which they derive their true meaning.

In line with its mission of cultivating global intelligence, the IKM doctoral program will develop intercultural skills and talents such as:

(1) Superior Intercultural Linguistic and Communication Abilities

In addition to English, which, for practical reasons, will be the lingua franca of the program, students will undertake an in-depth comparative study of at least two of the other principal languages of the world, in their cultural and intercultural context. These languages include, but are not necessarily limited to: Mandarin, Hindi, Spanish, Bengali, Arabic, Portuguese, Malay–Indonesian, Russian, Japanese, German, and French. If they are native speakers of any of these languages, they will choose two of the other principal languages, preferably those that are farthest removed from their mother tongue. For example, if they are native speakers of Hindi, they should not choose Bengali, but Mandarin and Russian; if they speak Portuguese, they should not choose Spanish or another Romance language, but Hindi and German, and so forth. This will ensure that students gain full access to linguistic and cultural worlds that are completely unfamiliar to them, so that their level of intercultural and linguistic understanding and, therefore, intercultural communicative skills, will eventually become even higher.

One should stress the fact that the program does not intend to train linguists or polyglots, any more than it intends to train political scientists, economists, lawyers, humanists, or any other specialists or experts.

Ideally, students who come into the program will already have genuine fluency in some of these languages (see admission requirements below). In-depth knowledge of a number of languages, however, is essential for the IKM student to feel at home in several cultures, move freely among them, and thereby gain a genuine global, cross-cultural perspective. Language courses will be taught in an intercultural comparative context so that students will become aware of the deep interconnections between the native speakers' linguistic and cultural worlds, including their fundamental systems of values and beliefs, religion, social, economic, and political behavior, historical development, civil institutions, and so forth. Language courses will also be taught in the context of the students' concrete research projects so that they will maximize the students' ability to carry out these projects.

(2) **Increased Intellectual Mobility and Flexibility**

The transdisciplinary and cross-cultural nature of the IKM research projects will require that students move between institutions in several regions of the world, as well as across departmental divides at any single institution. This kind of mobility will provide students with a local–global perspective, that is, with the ability to view a certain discipline or academic culture from both the inside and the outside. They will become immersed in the local research culture of a certain discipline or institution, at the same time that they will be able to reflect on it, by comparing it with other such research cultures. They will learn how to discern similarities and differences between them, which as a rule remain hidden to a partial, local view, as well as how to establish new links among them. A local–global perspective will give them the intercultural responsive understanding and flexibility needed to bring together specialists or experts from various fields and from several cultures in order to design and execute transdisciplinary and intercultural projects that none of these experts would be able to implement on their own.

(3) **Cross-Cultural Insight and Sensitivity**

The IKM doctoral program will create group solidarity among a culturally diverse body of students and will teach them how to cooperate in, and effectively interact with, shifting cultural and linguistic environments. By working together on intercultural and transdisciplinary projects, students will become aware of their different cultural assumptions in approaching a certain problem and will start negotiating among themselves to find the best solutions that go beyond their own local perspective or self-interest and advance the research project as a whole. Cross-cultural insight and sensitivity will also emerge from the daily interaction of students who will live, work, and play together as a group for an extended period and will be asked to build and act on a common sense of purpose and a common set of values for the rest of their lives. In other words, the students will be called on to seek global intelligence not only in relation to their academic studies, but also in their daily interactions both inside and outside their group. This kind of learning objective will, again, distinguish the IKM doctoral program from other foreign studies or study abroad programs that currently flourish all over the world.

(4) Ability to Integrate Academic and Experiential Knowledge

Global intelligence presupposes that students, from the first year of their studies, begin to acquire and combine theoretical and practical knowledge in order to address real-time, local–global issues. This learning objective will again distinguish the IKM program from current standard academic programs. These programs mostly convey an abstract body of knowledge, which is often disconnected from its practical, live context and which the student is supposed to apply or make use of at a later date, after graduation. By contrast, IKM students will organize their curricula and research programs around the concrete problems they are asked to solve, rather than solely on past case studies. They will form cross-disciplinary teams and work on viable solutions to specific real-world problems, rather than through the codified practice of a particular academic discipline or culture.

Students will, moreover, build capacity to identify and address potential socioeconomic and other types of problems before they develop into crises that threaten the peaceful development of world communities or diminish the diversity of world resources. They will also be called on to design workable, realistic blueprints for the sustainable, sociocultural and human development of their countries or regions. These blueprints will be based on the best traditions of wisdom available in their cultures, as well as in those of others, and on the most cherished aspirations and ideals of their people. Last, but not least, the sustained, cooperative efforts of the IKM graduates from all over the world will decisively contribute to addressing and eventually eliminating the causes of international terrorism, one of the greatest threats to humanity in our time.

6 Selection Criteria for Admission and Profile of Successful IKM Candidate

The entire success of the program will undoubtedly depend on its ability to attract high-quality applicants from all over the world and on a careful and rigorous selection process. It is therefore very important that the program offer substantial grants that will cover students' tuition, fees, travel, and living expenses, partially or entirely, for the whole period of study. This will not only ensure an outstanding pool of candidates, but will also avoid price discrimination, so that needy but very promising young men and women from all over the world can also apply and be admitted.

Applicants for the program can be recruited through extensive publicity in the international mass media and on the Internet, as well as through a nominating committee in each of the participating countries. This nominating committee would consist of prominent thinkers, educators, business executives, artists, writers, economists, community representatives, respected public and private figures, and so on. Potential candidates can also be brought to the attention of the admissions committee by colleges and universities, scholarship-granting foundations, professional and business associations, arts organizations, and by the students themselves in cooperation with their institutions.

The final selection will be made by the admissions committee, based on the recommendations of the C & R committee of the IKM program.

The selection process will consist of a preliminary and a final phase. Minimum requirements for admission to the preliminary phase include:

- An advanced degree (at least a master's or a professional degree) or equivalent in any field related to intercultural knowledge management. Fields include, but are not limited to, humanities, social and life sciences, computing, economics, international business and finance, law, environmental sciences, agriculture, architecture, civil engineering, medicine, pharmacy, and so on.

- Fluency in at least three languages, one of which must be English and one from a language group other than the candidate's native language.
- Superior communication and computer skills.
- Teamwork capacity.
- Proven physical and emotional ability to live and learn in a number of foreign cultures and ethnically diverse environments.
- Considerable work experience outside formal schooling.
- A substantial research proposal on issues related to intercultural knowledge management.

Candidates who advance beyond the preliminary phase will be invited to attend a four-week introductory summer course of study in intercultural knowledge management. The final selection for the program will be made shortly after the completion of the course.

The profiles of successful candidates include superior intellectual, linguistic, and communicative capabilities, proven creativity, proven ability to think and to relate to others in cross-disciplinary and intercultural contexts, and high personal integrity. Field of specialization will be less important than the candidate's willingness and ability to work cooperatively with specialists in all fields to carry out intercultural and cross-disciplinary projects. The most important quality of this profile will be a candidate's propensity toward global intelligence, that is, his or her ability and willingness to engage in intercultural responsive understanding and action on a global level, while never losing sight of the various local reference frames. Therefore, the introductory summer course in intercultural knowledge management will be indispensable for the final selection of the candidates.

7 Introductory Summer Course

This biannual, four-week summer session will take place in one of the IKM local centers, during June–July in the Northern Hemisphere, and January–February, in the Southern Hemisphere. The summer curriculum may include, but not be limited to: (1) Introductory course in intercultural knowledge management (definition of field, guiding theoretical assumptions, methodologies, general areas of research, etc.); (2) Workshop in intercultural communication and understanding; (3) Workshop in intercultural project management; (4) Workshop in comparative systems of values and beliefs; and (5) Workshop in advanced computer research technology platforms.

The session will serve to acquaint prospective students with each other and with the core faculty and other guest instructors. It will give an opportunity to candidates of diverse cultural and intellectual backgrounds to work cooperatively and to begin developing their study and research programs for their first year in the program.

In turn, members of the IKM core faculty and their associates will have an opportunity to interact with the finalists, as well as to observe and evaluate the candidates' interaction with each other. This will ensure that only the candidates who exhibit the qualities desired, above all a clear potential for global intelligence, will be finally admitted to the program. The unsuccessful candidates, who will undoubtedly be quite accomplished young men and women in their own right, will be given a certificate of attendance and graduate credit from the Institute. They will also be directed to other programs at the Consortium's host universities or other associated academic institutions that might better serve their career and life goals. In turn, the successful candidates will be given graduate credit for the summer program that will count toward their IKM doctoral degree.

8 Size of IKM Doctoral Program

The IKM doctoral program will be a pilot, experimental program that will create a small number of local–global practitioners and leaders and, therefore, will necessarily be very selective. We should be openly and unashamedly elitist in this respect, if by an "elite" we mean a corps of leaders that generously dedicate their lives and talents to the benefit of their local–global communities and human development as a whole. We should impress again and again upon our students—above all through our own conduct—that the mission and the responsibility of an elite, all too often forgotten and only rarely practiced by our current world leaders, is to serve others, not ourselves.

The IKM program will initially serve around forty graduate students per year. It is anticipated that, as a result of an enthusiastic worldwide response, the number of applicants will grow exponentially within a short time span. One should, however, maintain the most rigorous academic standards for the program, as well as its experimental quality, by wisely limiting the number of students enrolled. When fully developed, the program will enroll no more than fifty students per year or one hundred and fifty students in a three-year cycle. It is hoped that other institutions will follow suit and will develop similar programs in close cooperation with the present one. Global intelligence will ultimately favor cooperation over competition in all fields of human endeavor, so that an extensive network of local–global learners and practitioners can be developed throughout the world. No single institution or organization will have the huge human and material resources required to develop this kind of global network on its own.

9 Core Faculty

The core faculty of the IKM doctoral program will primarily consist of outstanding teachers and researchers from the universities participating in the Consortium. Members of the core faculty will help form an ad hoc curriculum and research (C & R) committee that will work closely with individual students to identify their major academic advisors at both the host and associated universities and to develop flexible curricula that are best suited to their particular research interests. They will also help develop the IKM research programs. The core faculty may consist of twenty to thirty members in the developing stages of the program and may grow up to fifty members, once the program is fully developed.

It is crucial that IKM core faculty members undergo the same global learning experiences that the IKM students do and that they possess high intellectual flexibility and versatility, responsive understanding, and ability and willingness to work cooperatively across disciplines and cultures. Indeed, the most important quality that an IKM faculty member should display is the same propensity toward global intelligence that is required of the IKM doctoral candidate.

Therefore, the selection of the IKM core faculty must be as careful as that of the student participants in the program. An appropriate selection process must be put in place, so that the best available faculty from the host and other academic institutions will be chosen. Once the program is well established, it will become a great honor and privilege for any faculty member to be associated with it, so it will be somewhat easier to identify and attract suitable teachers. The most outstanding and promising IKM graduates will also be invited to teach in the program.

IKM faculty members will be appointed for a four-year renewable term, so that the core faculty can be periodically updated and refreshed. After the completion of their tenure, faculty members will return to their home departments, but will continue to be active in the IKM project.

In addition to student advising, members of the IKM core faculty will participate in the introductory summer sessions and will cooperate with distinguished resource persons from various countries and academic and nonacademic fields in designing and teaching new cross-disciplinary and cross-cultural courses, as well as in developing and carrying out intercultural research programs. They will be responsible for assuring the highest academic and research quality, as well as the highest ethical standards (grounded in the principles of global intelligence) of the IKM program at all times.

During the initial phases of the IKM program development, there will be four annual workshops with members of the core faculty and resource persons from partner institutions and other academic and nonacademic organizations from around the world in order to determine the general area(s) and nature of the research projects and to design the required courses for the first and second academic years of the program. Each of these workshops will take place in a different part of the world and will be hosted by a local IKM center, in cooperation with the local IKM Consortium partner. Once the IKM program is well established, the number of the curriculum and research development workshops may be reduced to three or two per year.

10 Guest Faculty, Researchers, and Resource Persons

In addition to the IKM core faculty, there will be a constantly updated pool of distinguished faculty and researchers, selected primarily from the international universities and research institutions associated with the program. They will be called on to team-teach some of the courses or co-lead some of the research projects in the IKM program. When appropriate, these scholars and researchers may be offered visiting or adjunct positions with the program.

The IKM program will also rely on a large, constantly updated, pool of distinguished practitioners, drawn from as wide a global setting as possible. When appropriate, some of these practitioners may also join the IKM faculty as visiting or adjunct appointees. The two pools (academic and nonacademic) will ensure that, during the course of their studies, students will come into contact with prominent scholars, teachers, artists, writers, scientists, politicians, business people, public servants, community representatives, and personalities of diverse religions and spiritual traditions from the Middle East, western, central and eastern Europe, the Indian subcontinent, Africa, central, east and south Asia, the Americas, and so forth.

11 The IKM Research Programs, Curriculum, and Internships

Recognizing that disciplinary, departmental structures lead to fragmentation of knowledge and increasingly narrow fields of specialization, universities in the United States and other countries currently offer many interdisciplinary programs. Such programs are largely designed to address the unwanted consequences of a veritable explosion of knowledge, especially in the so-called hard sciences. But most of them bring together only neighboring or closely related disciplines and do not cut across the whole range of knowledge. In other words, they do not employ a transdisciplinary perspective, let alone an intercultural, global one.

This is true of most programs that may be regarded as more or less related to intercultural knowledge management, such as programs in international business, international relations, and global studies, as well as other programs in political science or sociology that focus on geopolitics, international trade, law, war and peace, conflict resolution, and so forth. Likewise, most international programs that deal with health, food, nutrition, population, and the environment, to say nothing of hardcore scientific research, only rarely take into account the vast differences in research cultures that can be observed in different parts of the world.

Research in most of these fields is oftentimes carried out in a piecemeal, isolated, monodisciplinary and monocultural fashion, without an attempt to integrate the various disciplines, such as history, religion, cultural anthropology, political science, economics, sociology, cognitive psychology and artificial intelligence, computing, environmental sciences, humanities, history and philosophy of science, and so forth, into a coherent theoretical framework and research programs that can adequately deal with complex human problems in a local–global reference frame.

Comprehensive and viable solutions to such problems can come only from a sustained, transdisciplinary and cross-cultural effort inspired by global intelligence or intercultural responsive understanding and action.

The IKM pilot project itself will attempt to reflect the emergent nature of global intelligence, by treating knowledge as a continuously emerging phenomenon at many different levels and within many reference frames. This is in marked contrast to the modus operandi of disciplinary knowledge or a constituted body of accumulated information, periodically updated and only seldom entirely restructured through so-called "scientific revolutions" (Kuhn 1970). Thus, the IKM project will create the adequate theoretical and institutional frameworks that will allow new knowledge to emerge and be continuously remapped and reorganized. The coherence and continuity of the IKM doctoral program will be ensured by unity of vision—global intelligence—rather than by unity of subject matter, which will be in perpetual flux.

In order to achieve its learning objectives, the IKM program will adopt creative ways of building curricula, flexible enough to allow students, in consultation with their academic advisors, to design their own course of study that will cut across various disciplines. The research program of each individual student will in part determine the content of his or her curriculum and practicum.

In addition to a number of required courses, designed to break down barriers between disciplines and to approach knowledge from a global perspective, students will do coursework relevant to their individual and collective research programs and will engage in concrete, real-time individual and group research projects on and off campus, working with appropriate advisors in various parts of the world, either in person or long-distance.

Finally, students will complete a number of internships with international, public and private organizations. These internships, no less than the curricula, will be fully integrated with the individual student's research program. Indeed, they will become important components of this research program.

The IKM Research Program

The IKM program will develop, on a yearly basis, concrete, real-time research projects in six, interrelated, general areas that are and will remain crucial for intercultural knowledge management in the foreseeable future:

(A) Globalization and strategies for human development

(B) Food, nutrition, and healthcare in a global environment

(C) Energy world watch and environmental studies for sustainable development

(D) World population movement and growth

(E) Information technology, new media, and intercultural communication

(F) World traditions of wisdom and their relevance to further human development

In consultation with other IKM adjunct faculty, researchers, and student representatives at host institutions from around the world, the C & R committee will decide which general area or areas and what kind of research projects within these areas might be accorded priority in a given year. Such priority will be based on the topicality, urgency, and global importance of the projects. In reaching a final decision, the C & R committee will also consider the research project proposals submitted by the IKM candidates for admission to the program during that year. Based on the merit of those proposals, it will make appropriate recommendations to the admissions committee as well.

During their first year with the program, students join small intercultural and cross-disciplinary research teams and work under a project leader or coleaders. The leaders are normally chosen from the IKM faculty. After their first year with the program, however, students can also become leaders or coleaders of their research teams.

Some research projects can be completed during relatively short periods, from several months to a year, and others may necessitate the sustained, collective effort of several years on the part of multiple, cross-disciplinary and intercultural teams. Each student will work on at least one of these larger research projects, out of which his or her doctoral thesis will emerge. Each student will work on concrete projects in all six general areas, so that s/he can develop an understanding of the basic issues and methodologies specific to those areas and learn how to establish new connections among them.

Below are listed a few samples of research projects that are of real-time global relevance and would contribute to solving important local and world problems. Needless to say, this list is only a model of the kind of projects that IKM students could work on. The actual, short-term and long-term, collective research projects will be negotiated and decided on by the IKM institutional partners.

1 Urban Sprawl: Cultural and Environmental Consequences. Global cities to be studied: Los Angeles, São Paulo, and Shanghai. Concrete recommendations to alleviate the situation will be proposed by a cross-disciplinary team of IKM researchers working closely with local officials, city planners, architects, ecologists, biologists, meteorologists, health and social workers, sociologists, political scientists, humanists, computer engineers, city resident committees, etc.

2 World Encyclopedia of Cultural Concepts. This project will involve interdisciplinary and cross-cultural research teams of linguists, semioticians, cultural anthropologists, sociologists, psychologists, philosophers, political scientists, economists, computer scientists, humanists, historians, artists, etc. The encyclopedia would not be eclectic, but would present each major cultural term in its historical perspective and in the context of a comparative analysis and cross-referencing with another similar or related term in a different culture or system of values and beliefs. The encyclopedia will at first focus on the cultural terms and usages in the larger world cultures and then will work toward including all other known cultures on our planet. This is a long-term project that will involve sustained research on the part of several generations of IKM students, faculty, and associated researchers from all over the world, as well as innovative use of high-performance data mining and management systems, based on artificial intelligence (AI) computing methods. Once the world encyclopedia reaches an advanced stage (although it will never reach a "definitive" form, being continuously updated and refined), it will be posted on-line and will become a basic reference work for any IKM student and local–global practitioner.

3 Local Blueprints for Socioeconomic, Cultural, and Human Development. This series of projects will research and propose middle- and long-range strategies for the socioeconomic, cultural, and human development of a certain country or region, based on a global and comprehensive analysis of its history, cultural traditions, political institutions, past and present social and economic performance, systems of values and beliefs, education, religion, relations with neighboring countries and regions, etc.

4 Global Health Project. This project will utilize high-performance management and data processing systems, based on AI computing methods, to work out the intercultural logistics and propose the implementation of a fully automated and interactive on-line global medical diagnostic and genetic data center. This global center will gather, process, and redistribute knowledge produced through the treatment of millions of patients and genetic information generated by thousands of research institutes and laboratories throughout the world.

The project will require the sustained, long-term effort of intercultural teams of researchers in fields such as medicine and pharmacology, genetics, bioinformatics, data management systems, social science, ethics, statistics, health care, intercultural studies, environmental sciences, and so forth, to assist in collecting, evaluating, and organizing medical and genetic data from doctor's offices, hospitals, genetic research institutes and laboratories, medical libraries, national health records offices, and so on, from all over the world.

The data and methods of collection and analysis will be based not only on the assumptions and practices of "allopathic," Western medicine, but also on those of the major schools of so-called alternative medicine. This global medical diagnostic and genetic data center will provide patients from all over the world with the most effective, comprehensive, and integrated treatments, will greatly advance medical and genetic research, and will detect and address outbreaks of epidemics and/or bioterrorist attacks at their incipient stages.

IKM Curriculum

The specific nature and content of the basic courses required for all students will be discussed and decided upon, on a yearly basis, by the ad hoc C & R committee. Once the program becomes fully operative, student representatives, elected by their peers, will also join the C & R committee. Just as in the case of the IKM research projects, the basic required courses proposed for each year will be a result of intercultural and interdisciplinary negotiations. The object of these negotiations will be to establish research priorities that will in turn determine what kind of course offerings might be appropriate for that year at different participating institutions from around the world.

The same input will guide the student selection process. Some courses, however, may be continued for a number of years, such as the language and intercultural communication and understanding courses, as well as the introductory courses in intercultural knowledge management. The latter courses will deal mainly with the basic theoretical questions and methodologies that are proper to IKM and will also be taught at the summer sessions with the finalists for admission to the IKM doctoral program.

The C & R committee may also publish on the Internet an annual list of available courses related to IKM and its research programs and offered during that particular year in various departments and schools of the host institutions and other IKM partners from all over the world. This list may be an important resource for incoming IKM students and their advisors and may assist them in designing their individual curricula. The final decision of what an individual curriculum might look like, however, rests with the individual student and his or her major advisor(s).

All courses are meant to provide methodological support for the students' research programs and may, in turn, provide ideas for such research programs. Many of the required courses will preferably be team-taught by an intercultural team of instructors from various host institutions. Another appropriate model would be to have course directors put together teams of guest lecturers from various outstanding academic and nonacademic institutions from around the world. Some courses could also have an electronic component, with guest lecturers appearing and interacting with their student audience on live teleconferences.

One can think of a number of course topics that will support research and that cut across a large number of academic disciplines, as well as across the six general areas identified in the preceding section on the IKM research programs (globalization and strategies for human development; food, nutrition, and healthcare; energy and sustainable development; world population movement and growth; information and communication technology; world traditions of wisdom). I have placed these topics under nine headings simply for the sake of convenience. Other headings can be added, or a different classification of relevant topics can be used. It is essential to keep in mind, however, that under any IKM classification system, all courses should be treated as cross-disciplinary and should, therefore, be cross-listed in an actual IKM curriculum.

I. *Theory and Practice of Intercultural Knowledge Management*

1. Intercultural Knowledge Management: Theory and Practice (Introductory Course)
2. Intercultural Research and Learning Technology Platforms: Principles, Methods, and Practice
3. Series of Workshops in Intercultural Project Management
4. Globalization and Local Cultural Heritage
5. Creative Thinking and Action: Culturally Productive Metaphors in Their Local and Global Contexts

II. *The Nation–State and Local, Regional, and Global Communities*

1. National Identity and Sovereignty: Historical and Theoretical Approaches. The Role of the Nation–State in Global Societies
2. Series of Workshops and Seminars on Various Regions and Cultures of the World
3. Border Cities and Regions As Intercultural Focal Points
4. Regional and Global Political Organizations (The United Nations, The League of Nations, NATO, etc.): History, Objectives, Structure, and Performance
5. Modernity and Postmodernity in Local and Global Contemporary Discourse and Practice

III. *Global Markets, Finance, and the World Economies*

1. History of Trade Practices from Around the World
2. History and Future of Money and Financial Practices
3. Global Policy Decision Making: Trade, Global Markets, and International Organizations and Agreements
4. Toward an Ecology of World Commerce

IV. *Sustainable Development and the Future of Humanity*

1. World Disarmament and Conventional Weapons
2. World Population, Immigration, and Sociocultural Displacement
3. Climate Changes, Natural Environment, and the World Economies

4. Mobility, Urban Planning, and Clean Technologies
5. Public Health and Global Water and Land Management
6. Natural Capitalism: Theory and Practice

V. *World Systems of Values and Beliefs*

1. Seminar Series in Comparative World Ethics
2. Seminar Series in Comparative World Religions
3. The Human and the Nonhuman: Concepts of Humanity in Various Cultures and Disciplines
4. Traditions of Wisdom and Their Contemporary Relevance

VI. *Violence and Human Societies: Origins, Causes, and Remedies*

1. Violence, Religion, and Culture
2. Environment, Natural Resources, and Violence
3. Violent Conflict and Conflict Resolution in the 21st Century
4. Crime and Punishment: Ideas of Justice and Law Enforcement in Various Cultures from Antiquity to the Present
5. International Terrorism: History, Causes, and Prevention

VII. *Science, Technology, and Culture*

1. Social and Ethical Implications of Science and Technology
2. Scientific Fashions and Revolutions: A Historical Overview
3. World History of Technology and Technological Development
4. Annual Colloquium on the State of Knowledge in the Natural and Human Sciences

VIII. *Information Technology and the New Media*

1. Social and Ethical Implications of Information Technology and New Media
2. Information and Communication Technology and Strategies for Human Development
3. The New Media in an Intercultural Environment: Global Mission and Responsibilities
4. Information Technology, Intellectual Property, and Intercultural Exchange

IX. *Language, Cognition, Interpretation, and Communication*

1. Series of Intercultural Workshops on Linguistic Communication and Understanding
2. English As a Global Language: The Tower of Babel Reconstructed?
3. Interpretation, Communication, and Cultural Translatability
4. Language, Gender, and Culture
5. Transcultural Laboratory

Each of these courses will require cross-disciplinary readings in several languages and from several cultures. Readings may include, but not be limited to historical, sociological, anthropological, economic, scientific, philosophical, psychological, religious, and literary and other artistic works. Some courses may also require electronic video and motion picture presentations, as well as audition of, or even participation in, live art performances.

Internships

During their IKM studies, students will complete at least three internships, for a two-month period each: the first internship will be with an international nongovernmental organization, the second one with a governmental institution, and the third one with a multinational corporation. The purpose of these internships is for students to learn and understand from first-hand experience how these organizations operate, to what purpose, and with what success. Internships will, moreover, be directly related to the individual student's research programs and are designed to advance these programs. For example, if the student researches urban sprawl, he or she might take an internship with the city planning office of Los Angeles or Shanghai or Mumbai; if he or she studies global health issues, he or she may take an internship with a global health governmental or nongovernmental organization, environmental agency, and so forth.

Upon completion of the internship, students will submit a position paper on their learning experience, including a set of proposals designed to improve the performance of that particular international organization (in the spirit of global intelligence). All internships will carry academic credit.

12 Advanced Learning and Research Technology Platform and IKM Online

The IKM project will not be a distance learning organization, even though it will certainly cooperate with such organizations and will greatly benefit from their experience and services. The IKM project is based on the belief that, in addition to global expertise, students should strive for global intelligence. Such global intelligence cannot be acquired through electronic means or virtual reality, but only through the students' real-time intercultural experience and practice of living, working, and playing together over extended periods in various cultures and learning environments. The IKM project also starts from the premise that information and communication technology, as well as technology in general, is only as good (or as bad) as the intentions and objectives of the people who use it.

The IKM project will, therefore, leverage information technology (1) to achieve a closely knit, global community of students, teachers, scholars, and practitioners who support each other's research and learning efforts toward acquiring global intelligence, and (2) to disseminate the results of these learning efforts to the world community at large.

In order to attain these objectives, the IKM project will form partnerships with appropriate leading-edge IT and distance learning companies in order to develop innovative, learning and research technology platforms. Such technological platforms are needed to offer effective global research tools to the IKM students, mentors, and associates from all over the world. The program's data management system, based on AI methods, must be capable of analyzing vast and complex amounts of data, of recognizing cohesions and suggesting intelligent, optimized decisions. The system will provide real-time, forecasting strategies and will be used to help model and solve comprehensive scientific, social, economic, and cultural problems. It will, therefore, be an important tool in carrying out the IKM research programs.

The IKM program will also develop a teleconference and video production lab (TVPL), which will serve as the electronic, web-based link among the various IKM local centers around the world. Through the TVPL, the program will offer further opportunities for distance learning and will introduce multimedia technologies in IKM classrooms all over the globe.

13 IKM Program Schedule and Logistics

The IKM doctoral program will be a three-year, full-time, study program. In order to attain its learning objectives, the IKM project requires that students divide their time between several locations in various parts of the world, carrying out research, taking courses, and completing internships related to this research. They will begin their intercultural immersion even before their first year with the IKM program, during the introductory summer sessions. Once admitted to the program, students will typically spend their first term in a local center of intercultural learning and research, on the main campus of a host university, for example, in the United States. If required by their individual or group research, however, they will also undertake up to three-week study trips within the United States and the geographical regions that are most accessible from North America, such as Canada, Central America, or the Carribean Islands, where they will be hosted by the IKM project associates in the region.

During their second term, students will be stationed in Europe, for example in the South of France or Northern Spain, with up to three-week study trips, if required by their research, within Europe, the Middle East, and North Africa. IKM local partner institutions may host them in these regions as well. During their third term, students may go to a local center in Russia, China, or India, as required by the specific objectives of their group research and/or internship.

Depending on their research programs, however, students may, in consultation with their advisors and subject to approval by the C & R committee, propose different overseas research schedules during their second and third years with the IKM program.

During the last semester of their third year, students will return to and work at the local center that will be most appropriate for their main collective research projects and doctoral dissertations. They will defend their doctoral theses and receive their Ph.D. degree at the end of their third year of full-time study.

14 IKM Graduate Employment Opportunities and the IKM Global Network

There is no doubt that every national and transnational public and private entity throughout the world will seek to hire graduates from a doctoral program in intercultural knowledge management. IKM graduates will secure high-level jobs with transnational corporations, national and international governmental and nongovernmental institutions, media organizations, universities, foundations and think tanks, or will set up their own consulting companies and transnational firms. Some of the most promising graduates will also find employment with the IKM doctoral program, which will continually need to train not only its students but also its teachers.

Once they receive their doctoral degrees, the IKM graduates will disperse to many regions of the world. The spirit of global intelligence requires, however, that they continue maintaining close contact and working together toward the same goals and ideals for the rest of their lives. To ensure this group solidarity, all IKM graduates will be granted the status of permanent fellows of the IKM Institute.

The local centers will organize annual gatherings for new and old IKM fellows from various parts of the world. Many of the IKM fellows will be invited to teach, lecture, conduct workshops, lead, or colead new research groups and projects in the IKM doctoral program. They will also become involved in recruiting new students for the program.

Over time, the IKM network can provide governmental and nongovernmental agencies, the international media, and other organizations with a roster of local–global practitioners in numerous fields. IKM fellows may well serve as point men and women in interagency and international negotiations and conflict resolution. Indeed, many of them will eventually be in a position to negotiate national and regional policy, global and local interests, global and local economic and financial investments, and common strategies of cultural and human development on behalf of governments and institutions they may head or represent in various parts of the world.

15 Financial Considerations

Potential financial backers for the project will certainly wonder about the cost of a global educational program of this type. Such financial backers may include the members of the IKM academic consortium itself, various private national and international foundations and other charitable organizations, governmental and intergovernmental development grant agencies, individual donors, and so forth. The initial financial investment in this global learning experiment would undoubtedly be substantial: one would have to build and continuously update the complex infrastructure and the ICT needed for the program. One would also have to cover the educational, travel, and living expenses of all IKM students (up to fifty enrollments per year), as well as the salaries and travel expenses of the core faculty and other resource persons.

The costs, however, will not exceed those needed to train a physician, lawyer, or an MBA graduate in the United States, and the benefits to the global society at large would obviously be much greater. Likewise, only a very small fraction of what is currently spent in the United States and other countries on the so-called war on terror and other military adventures (so far with very uncertain results) would suffice to create a large number of IKM programs all over the world. Such programs would, moreover, yield much better and more secure returns, for both the United States and the rest of the world. So, even in terms of utilitarian benefits or returns, the IKM project would not be a bad investment. Indeed, it would become self-supporting after the first three-year cycle, because of its real-time, advanced research programs that many multinational corporations and other transnational, private, and public organizations would regard as very hot intellectual property.

At any rate, the intrinsic value of the program to the global community at large will greatly exceed any financial investment needed to establish and operate it. One certainly should not ignore cost/benefit considerations. Yet, one would again have to redefine the notions of "value" and "benefit," not in the utilitarian, instrumental terms of material self-interest, but in terms of the goals and objectives of an emergent ethics of global intelligence. In the end, it is a matter of choice on the part of a certain society or community as to what its investing priorities should be. Will it continue indulging in mindless waste of human, natural, and financial resources, with disastrous, worldwide repercussions? Or will it finally start building a sustainable future for itself and for all other life on earth?

Notes

Introduction

1. See Michael Hardt and Antonio Negri, *Empire* (Cambridge, Mass: Harvard University Press, 2000, pp. 213–218). That Hardt and Negri's book has been hailed in U.S. academic postmodernist circles as the new *Communist Manifesto* speaks, I believe, for itself. I shall again refer to this study in part III of the present book. See also my extensive discussion of it in *Remapping Knowledge: Intercultural Studies for a Global Age*, which is a companion volume to the present one.

2. For pleasure and suffering as physiological manifestations of the will to power, see Friedrich Nietzsche, *The Will to Power* 1967 [1901], especially pp. 369–375. For example, Nietzsche asserts: "If the innermost essence of being is will to power, if pleasure is every increase of power, displeasure every feeling of not being able to resist or dominate; may we not posit pleasure and displeasure as cardinal facts?" (ibid., p. 369).

In order to create a genuinely new ethics, Singer would first need to redefine pleasure and suffering in terms other than the will to power, which would also mean redefining them in terms other than utilitarianism. His theory of "rights," also of utilitarian extraction, may be of some use in current political activism concerned with the protection of animals, but is woefully inadequate as an emergent ethical standard of global intelligence. For an excellent critique of utilitarianism and rights theory from an ecological perspective, see Gare 1993a, pp. 44–51.

3. Arran Gare evaluates the view of "mechanical humans" or human machines, corresponding to "mechanical nature," as a triumph of Western "nihilism." He traces this view back to Hobbes to whom he also ascribes, just like Singer, the concept of humans as motivated primarily by power and material self-interest. For Gare, moreover, this view has "underlain the fields of politics, economics and psychology," so that the "modern doctrines of rights theory, utilitarianism, mainstream economic theory, Social Darwinism and behaviorist psychology are simply the working out of this [Hobbes's] research program" (Gare 1993a, p. 135). Needless to say, Gare, in contradistinction to Singer (and to some extent to Dyson, who dedicates the first chapter of his book to Hobbes) does not

attempt to redeem this view. On the contrary, he sees it as being at the root of most of the Western-induced, environmental and other crises that confront humanity today. I shall further discuss Gare's post-Marxist environmentalism in chapter 5, below.

4. In my earlier work (Spariosu 1997), I have called this alternative system of values and beliefs an *irenic mentality*, from the Hellenic word for "peace." I have also argued that irenic mentalities are incommensurable in relation to the various mentalities of power currently prevailing on our planet. I shall further develop these concepts and distinctions throughout the present study.

Chapter 1

1. There is an ever-growing literature on globalization, especially on its political and economic dimensions. Most of the approaches to this elusive phenomenon (whose novelty, or even existence, is barely recognized by a large section of the scholarly community) are necessarily interdisciplinary, involving such fields as sociology, political science, international relations, political geography, economics and finance, information and communication technology, the new media, etc. Prominent scholars and practitioners such as Janet L. Abu-Lughod (1989), Benjamin R. Barber (1995), Pierre Bourdieu (1998), Manuel Castells (1996, 1997, 1998), Thomas Friedman (2000), Francis Fukuyama (1992), Arran Gare (1993a, b), Anthony Giddens (1999), John Gray (1999), Michael Hardt and Antonio Negri (2000), David Held (1995), Samuel Huntington (1996), Kenneth Jowitt (1993), Scott Lash and John Urry (1988), Ervin Laszlo (1974, 1997), Marshall McLuhan and Quentin Fiore (1968), George Soros (1998), Joseph Stiglitz (2001), Immanuel Wallerstein (1991, 1999), and many others engage topics ranging from world systems theory, sovereignty, and the nation–state to global corporate and political governance; from economic liberalism and free trade to deterritorialization of transnational corporations and financial markets; from international security to global nuclear and chemical disarmament; from local ethnic and religious warfare to worldwide terrorism; from regional environmentalism to global sustainability; from health and population concerns to mass migration, diasporic communities, and human and animal rights; from transnational networks of nongovernmental, civic, religious, and women's organizations to global civil society and cosmopolitan democracy; from information technology and the global village to the new global media and the digital age, and so on.

There are also a significant number of studies that focus on the cultural dimensions of globalization, but most of them equally subordinate these dimensions to economics and politics. The main topics here are the human consequences of globalization (as the last stage of modernization); transnational cultural capital and the new media; globalization and popular culture; postcolonialism and identity politics; global culture and gender wars; global environmental crisis, and so forth. The field is dominated mostly, but not entirely, by Marxist, post-Marxist, feminist, poststructuralist, and other social critics housed in academic departments such as English, Comparative Literature, Women's Studies, Film Studies, and Cultural Studies, and, occasionally, in Cultural Anthropology, Environmental Science, History, Political Science, Philosophy, Sociology, and History of Science.

Among the studies that focus on the cultural aspects of globalization from various theoretical perspectives, I have largely drawn on the following: Mehdi Abedi and Michael Fischer, *Debating Muslims: Cultural Dialogues in Postmodernity and Tradition* (1990); Arjun Appadurai, *Modernity at Large: Cultural Dimensions of Globalization* (1996); Zygmunt Bauman, *Globalization: The Human Consequences* (2000); Peter L. Berger and Samuel P. Huntington, editors, *Many Globalizations: Cultural Diversity in the Contemporary World* (2002); Terry Eagleton, *The Idea of Culture* (2000); Ann Cvetkovich and Douglas Kellner, editors, *Articulating the Global and the Local* (1997); Mike Featherstone, *Undoing Culture: Globalization, Postmodernism, and Identity* (1995); Mike Featherstone, Scott Lash, and Roland Robertson, editors, *Global Modernities: From Modernism to Hypermodernism and Beyond* (1995); Lawrence E. Harrison and Samuel P. Huntington, editors, *Culture Matters: How Values Shape Human Progress* (2000); Samuel P. Huntington, *The Clash of Civilizations and the Remaking of the World Order* (1996); Fredric Jameson and Masao Miyoshi, editors, *The Cultures of Globalization* (1998); John F. Kavanaugh, *Following Christ in a Consumer Society: The Spirituality of Cultural Resistance* (1991); Anthony D. King, editor, *Culture, Globalization and the World System: Contemporary Conditions for the Representation of Identity* (1997); Joseph Nye, *Soft Power: The Means to Success in World Politics* (2004); Roland Robertson, *Globalization: Social Theory and Global Culture* (1992); Janusz Symonides, Kishore Singh, et al., *From a Culture of Violence to a Culture of Peace* (1996); John Tomlinson, *Globalization and Culture* (1999); and Immanuel Wallerstein, *Geopolitics and Geoculture: Essays on the Changing World-System* (1991). See also Ruy O. Costa's comprehensive paper, "Globalization and Culture," presented to the Advisory Committee on Social Witness Policy of the U.S. Presbyterian Church (2003).

2. Peter Sloterdijk in *Sphaeren I. Blasen* (1998) develops an interesting theory of the global as the expression of human desire for integration or completion. The globe or the sphere is the most extensive spatial representation of this desire, from Aristophanes' myth of the androgyne in Plato's *Symposium* to the various representations of the celestial spheres, to the human foetus in the womb. In my terminology, Sloterdijk's theory is a (proper) form of globalism.

3. For example, Ann Cvetkovich and Douglas Kellner argue that we need "to think through the relationship between the global and the local by observing how global forces influence and even structure ever more local situations and ever more strikingly. One should also see how local forces and situations mediate the global, inflecting global forces to diverse ends and conditions and producing unique configurations for thought and action in the contemporary world" (Cvetkovich and Kellner 1997, pp. 1–2).

In turn, Roland Robertson notes that in a global frame, the issue of universalism and particularism arises in terms of identity representation. He distinguishes between two theoretical positions: "relativism" and "worldism." Relativism, under which label he also includes postmodernism, refuses to "make any general, universalizing sense of the problems posed by sharp discontinuities between different forms of collective and individual life." It believes that speaking of culture in a global perspective "almost inevitably involves participation in a game of free-wheeling cultural politics." Consequently, relativism regards culture as being "inextricably bound up with power and resistance (or liberation)" (Robertson 1992, p. 99f).

By contrast, worldism claims that "it is possible and, indeed, desirable to grasp the world as a whole analytically," so that events happening around the globe can be interpreted in terms of the dynamics of the entire world system. This does not, however, preclude analyzing the representation of identity in terms of cultural politics, because "many of those who emphasize culture as a privileged area at the present time make diffuse, highly rhetorical claims as to its grounding in a world-systemic, economic realm" (ibid., p. 100)—a transparent allusion to Wallerstein's Marxist, heavily economic and sociological approach.

Robertson goes on to suggest that universalism and particularism "have become united in terms of the universality of the experience and, increasingly, the expectation of particularity, on the one hand, and the experience and, increasingly, the expectation of universality, on the other." The particularization of universalism means that the universal assumes "global–human concreteness." In turn, the universalization of particularism implies that "there is virtually no limit to particularity, to uniqueness, to difference, and to otherness" (ibid., p. 102). According to Robertson, even the "discontents of globality" operate along the "universalism–particularism axis of globalization" (ibid.). For example, radical Islam's resistance to globalization can be regarded as opposition not merely to the view of the world as one homogenized system—as Western modernism would have it—but also to the idea of the world as a series of culturally equal, relativized, entities or ways of life, as Western postmodernism would presumably see it.

4. Robertson also distinguishes five phases of world historical development that has led to the present "very high degree of global density and complexity" (Robertson 1992, pp. 58–59):

(1) *The Germinal Phase*, spanning between the early fifteenth and the mid-eighteenth century, in Europe. It includes globalizing tendencies such as the growth of national entities and the downplaying of the medieval "transnational" system; the expansion of the Catholic Church to various parts of the world; the emergence of heliocentric cosmography; the initiation of modern geography; and the spread of the Gregorian calendar. But, one may also mention such obvious developments as the European discovery and colonization of the Americas and the European colonization of large parts of Africa and Asia; the growth and peaking of the Ottoman empire; the Russian expansion into Asia; the establishment of the Austrian–Hungarian empire, etc.

(2) *The Incipient Phase*, spanning between the 1750s and the 1870s, again in Europe. This phase is characterized by the spread of the homogeneous, unitary nation–state; formalized international relations and an increased number of legal conventions and agencies concerned with international and transnational regulation and communication; international exhibitions; the question of admission of

non-European societies to the international "community"; the thematization of the nationalism–internationalism issue. But one may also mention such globally influential events as the French and the American Revolutions, the Napoleonic wars, the Latin American, African and Asian colonial wars, etc.
(3) *The Take-Off Phase*, spanning between 1870s and 1920s. It is during this phase that globalization assumes the tetradic framework proposed by Robertson: national societies, generic individuals (with a masculine bias), a single international society, and an increasingly singular, although not unified conception of humankind. Main globalizing tendencies during this phase include: the thematization of the issue of modernity; the thematization of national and personal identity; the inclusion of a number of non-European nations in the international "community"; increase in speed and number of global forms of communication and transportation (telephone, telegraph, radio, train, automobile, airplane, etc.); global competitions, such as the Olympics and the Nobel Prize; implementation of world time and near-global adoption of the Gregorian calendar; and World War I. One should not forget, however, the creation of various socialist and communist internationals, the Russian Revolution of October 1917, and the establishment of the multinational and multiethnic Soviet Union.
(4) *The Struggle-for-Hegemony Phase*, spanning between the mid-1920s and late 1960s. This phase is characterized by conflicting notions of modernity, leading to World War II and the cold war afterwards; the use of the atomic bomb and other weapons of mass destruction; the Holocaust; the establishment of the principle of national independence; the League of Nations and then the United Nations organization; the crystallization of the Third World. But one should also mention the massive population displacements caused by wars, revolutions, and sociopolitical, cultural, and religious repression all over the world.
(5) *The Uncertainty Phase*, spanning from the late 1960s to the present. Globalizing tendencies during this phase include: the American and Russian outer space programs, and the American landing on the Moon; the end of the cold war and increase in the fluidity of the international system; the sharp rise of the question of human rights and global civil societies; widespread access to nuclear and thermonuclear weapons; rise of fundamentalist religious movements; sharp acceleration in means of global communication and transportation; consolidation of the global media system; concern with humankind as a species community, especially through global environmental movements; conceptions of the individual rendered more complex and problematic by gender, sexual, ethnic, and racial factors; Earth Summit in Rio de Janeiro. But one should also mention, in addition to the "ethnic revolution" (ibid., p. 59), worldwide decolonization and increase in political freedom movements, as well as political terrorism, all over the planet.

By Roberston's own admission, this outline is to be fleshed out and "much detail and interpretation of the shifting relationships between and the relative autonomization of each of the four major components to be worked out" (ibid., p. 60).

5. By contrast, Janet L. Abu-Lughod (1989), for instance, offers a less Eurocentric picture of the "world system" than Robertson does, in her study

significantly entitled: *Before Hegemony: The World System* A.D. *1250–1350.* Abu–Lughod's intercultural historical approach to the late Middle Ages may be extended to other periods of world history as well.

6. See *The New Science of Giambattista Vico*, translated by Max H. Fisch and Thomas G. Bergin (Ithaca, N.Y.: Cornell University Press, 1968). For a fine study of Vico's poetic thought, including his cyclic notion of history (based, in my view, on the principle of mutual causality), see Giuseppe Mazzotta, *The New Map of the World: The Poetic Philosophy of Giambattista Vico* (Princeton, N.J.: Princeton University Press, 1999), especially chapter 9, "The *Ricorso*: A New Way of Seeing," pp. 206–233.

7. In *The Wreath of Wild Olive*, I have drawn a distinction between compatible, incompatible, commensurable, and incommensurable worlds (Spariosu 1997, pp. 58–66). This distinction can, in the present context, be applied to what systems theorists call "polysystems" (Even-Zohar 1990). Such polysystems can be composed of correlated systems, subsystems, and suprasystems. Compatible polysystems have similar organizing principles and kindred or easily adjustable reference frames. When they come into contact with each other, they tend to cooperate, rather than clash. In turn, incompatible polysystems have organizing principles and reference frames that are not easily interadjustable. When they come into contact, one of them will often annihilate or incorporate the other, or both will fuse into a different system. Commensurable polysystems are incompatible, but their organizing principles and reference frames are understandable and translatable into each other's terms. By contrast, incommensurable polysystems (or worlds) have organizing principles and reference frames that cannot be translated into each other's terms. The notion of liminality that I introduce later on in the argument will be useful in explaining how communications and passages might actually take place between incommensurable worlds. For further elaboration of these points, see chapter 2, as well as chapters 4 and 5 below.

8. Hans Vaihinger develops a functional theory of the relationship between fiction, hypothesis, and dogma in his monumental neo-Kantian treatise, *The Philosophy of "As If": A System of the Theoretical, Practical and Religious Fictions of Mankind*, translated by C. K. Ogden (London: Macmillan, 1924).

9. Of course, one can also find Western theorists who attempt to mediate between the two positions, or subject them to a Hegelian sublation (*Aufhebung*), but this approach has so far met with little success, being usually relegated to the first, universalist position. A good example of a universalist, but flexible and conciliatory view is that of Terry Eagleton (2000). Appadurai (1996), on the other hand, represents an intransigent version of the second view. For a full discussion of Appadurai's concept of conflictive difference and cultural politics, see my *Remapping Knowledge*, chapter 1.

10. I wrote a first version of this chapter a year before September 11, 2001. Since then Huntington's book has received a new lease on life, because the tragic events at the New York Trade Center and the Pentagon seem, at least in the view of some political leaders, to have validated its theses. Needless to say, I do not share this view. On the contrary, the events of September 11 and subsequent global-

itarian reactions to them emphasize, more than ever, the urgent need for working toward an ethics of global intelligence. It is only through cultivating a mentality of peace that humans can disengage from the destructive, amplifying feedback loops that violence invariably generates. On the other hand, irenic values should not be confused with those of pacificism, which largely remains within a mentality of power. See part II of the present study for further elaboration of these points.

Chapter 2

1. For an interesting essay on resonance in the context of reader response theory in literary studies, see Wai Chee Dimock, "A Theory of Resonance," *PMLA* (October 1997), vol. 112, no. 5, pp. 1061–1071. Dimock, however, is unaware of the relationship between mimesis and resonance in the Western literary tradition. Instead, she develops a distinction between resonance and dissonance in terms of a "democratic" conflict of textual interpretations. See also Kurt Spellmeyer's comments on resonance, in *Arts of Living* (2003). Spellmeyer points out that the modern term comes from the rediscovery of harmonics by Renaissance science. In turn, resonance in the world of culture "signifies the achievement of harmony—intellectual, emotional, aesthetic, and sensuous—between our small worlds and the larger one. The deeper our experience of resonance, the more encompassing the small world becomes until it seems to connect us with absolutely everything" (Spellmeyer 2003, p. 9). For Spellmeyer, resonance is the means through which culture, and the humanities in general, can "expand the small world outward" and "make a home of the universe" (ibid.). Like Dimock, however, Spellmeyer is unaware of the negative, amplifying feedback loops that resonance can produce in power-oriented cultures.

Finally, compare Spellmeyer's useful comments on the negative consequences of "criticism" and "critique" in the contemporary humanities, which he sees as a means of reproducing and perpetuating consumerist culture and which he proposes to replace with the terms "interpretation" and "creation" (ibid., pp. 145–173). He does not, however, see the intimate connection between criticism and crisis, and the key role of these concepts in perpetuating a mentality of power in general. Nor does he realize that "interpretation" and "creation" equally function as power instruments in Western culture. For a full argument, see my *Remapping Knowledge*, chapter 1.

2. See Mikhail Bakhtin, *The Dialogic Imagination: Four Essays*, edited by Michael Holquist and translated by Caryl Emerson and Michael Holquist (Austin: University of Texas Press, 1981), pp. 272–280. For a full discussion of Bakhtin's concept of responsive understanding and the ways in which it can be rethought in terms of a mentality of peace, see my study, *The Wreath of Wild Olive*, pp. 98–102 and passim.

3. I fully develop this argument, as well as the entire notion of a reconstructed field of intercultural studies in my *Remapping Knowledge*, chapters 1 and 2.

Chapter 3

1. For example, Wilson contends that although "it is true that science advances by reducing phenomena to their working elements—by dissecting brains into neurons, for example, and neurons into molecules—it does not aim to diminish the integrity of the whole. On the contrary, synthesis of the elements to re-create their original assembly is the other half of scientific procedure. In fact, it is the ultimate goal of science" (Wilson 1998, p. 234). As we shall see, Wilson's claim is not a mere historical slip, but a deliberate strategy of reducing all science to Western-style, mainstream science. For an extensive critique of Wilson's scientific rhetoric of power, see also my essay, "Sociobiology, Natural Science, and Human Development in a Global Age," in Aczel and Nemes (2003), pp. 205–228. The present discussion is a distilled and refined version of that essay.

2. For a full discussion of these issues, see my *God of Many Names: Play, Poetry and Power in Hellenic Thought from Homer to Aristotle* (Durham, N.C.: Duke University Press, 1991).

3. See Michael Loewe, *Chinese Ideas of Life and Death* (London: Allen and Unwin, 1982). Loewe shows that the idea of a purposeful, anthropomorphic creator is present in the Chinese classical tradition, for example in the Taoist notion of *tsao wu che* (p. 68). But this idea takes second place to the Taoist holistic and dynamic notion of the cosmos, in which the distinction between creator and creation becomes inoperative. Cf. my discussion of Taoism in chapter 4 below.

4. Even in contemporary biology, many scientists are uncomfortable with this idea. Instead, biologists now speak of group stability. For example, John Maynard Smith and other ethologists speak of "evolutionarily stable strategies" (ESS), where selection favors middle-of-the-road group behavior, penalizing "strong" deviant individuals. In turn, molecular geneticists and biochemists prefer to separate natural selection from its agonistic implications, regarding it simply as a differential rate in reproduction. Stephen Jay Gould and Niles Eldredge's theory of evolution as "punctuated equilibrium" shifts the emphasis from struggle for survival to random variation, contingency, and the play of chance. Finally, Wilson himself has in the past done fine scientific work on the evolutionary value of altruistic behavior and biodiversity.

See John Maynard Smith, *The Theory of Evolution* (Cambridge University Press, 1956; third edition, 1975); Stephen Jay Gould and Niles Eldredge, "Punctuated Equilibrium: The Tempo and Mode of Evolution Reconsidered," in *Paleobiology* 3 (1977); David Loye's essays on a "new global ethic" in *World Futures: The Journal of General Evolution* (2001), as well as his book, *Darwin's Lost Theory* (New York: Seven Stories Press, 2002); and Lynn Margulis and Dorion Sagan, *What Is Life?* (New York: Simon and Schuster, 1995). I shall return to the ideas of Gould, Eldredge, and Margulis in this chapter, below.

5. For a detailed analysis of Marlowe's views on power and globalization in *Tamburlaine the Great*, which can easily be extended to his *Dr. Faustus* as well, see my *Remapping Knowledge*, chapter 2.

6. Von Bertalanffy, a biologist and philosopher who belonged to the Vienna circle, started working on systems theory about the same time as and in parallel to the advent of quantum mechanics. Reflecting on his early work, he notes that the prevalent mechanistic approach in biology at that time "appeared to neglect or actively deny just what is essential in the phenomena of life," i.e., their holistic nature. To counter this mechanistic approach, he began to advocate "an organismic conception of biology which emphasizes consideration of the organism as a whole or system" and sees the main goal of the biological sciences "in the discovery of the principles of organization at its various levels" (von Bertalanffy 1968, p. 12).

7. As Magoroh Maruyama (1974), for instance, points out, the unidirectional causal paradigm presupposes the notions of substance and identity. In the Western philosophical tradition, beginning with Aristotle, reality has predominantly been seen in terms of entities or substances. Linear causation is a transfer of attributes from one substance to another, without a loss of identity on the part of the transmitter, which remains unchanged. In turn, to be an entity caused or transformed by another entity means to experience its force and receive some of its attributes. Linear causal relations appear as intermediate forces between two separate entities, for example between an agent and an agent's action. Such action does not affect the agent, moreover, but only the receiver.

8. But even scholars who, like Maruyama, embrace systems theory and nonlinear thinking remain unaware of their roots in ancient Western and other traditions. For example, Anatol Rapoport, in "A Philosophical View" (in Milsum 1968) writes: "The most fundamental difference between the modern and the ancient views of the world is in the emphasis on time processes in the modern view. To be sure, processes, in particular evolutionary ones, play an important part already in Aristotle's philosophy. However, the ancients did not have the proper analytical tools to construct a consistent and productive theory of such processes. The logic of classes and categories (the only rigorous analytical tools commanded by the ancients and by the schoolmen) is adequate for a static theory, which describes a portion of the world "as it is"; but it is not adequate for a dynamic theory, which is to describe the flux of interdependent events" (pp. 4–5).

Rapoport seems to believe, not unlike Wilson, that the "proper tools" for any serious scientific endeavor became available only in Europe and then only in the Age of Reason. He is plainly wrong, however, as my discussion of early Buddhism and Taoism in chapter 4 below will show. The fact is, as we have seen in the case of Wilson, that many Western scientists, even the most distinguished ones, do not have the necessary intercultural knowledge to make these comparative judgments. But ignorance does not prevent us Westerners from venturing such judgments anyway, seemingly because of the "heroic" hubris that the global success of our power-oriented endeavors breeds in us. Modesty, open-mindedness, and self-effacing humor might in the end be the proper intellectual attitudes for all of us scholars and practitioners in a global (and local) environment.

9. Initially, this notion was applied to self-regulatory mechanisms, such as Cornelius Drebbel's thermostatic control of automatic furnaces in the seventeenth century, or James Watt's governor on the steam engine in the eighteenth century, or Norbert Wiener's control system in servomechanisms of antiaircraft guns, during World War II. The output of these closed systems is monitored back to their receptors, in order to regulate their performance in relation to preset objectives. There are two types of feedback loops: negative feedback loops that stabilize the system within its preestablished trajectory, reducing deviation between goal and performance; and positive feedback loops that reinforce or amplify the deviations, thus producing instability in the system, which may eventually lead to the system's modifying its goals. For a brief history of these self-regulatory mechanical devices, see M. D. Rubin, "History of Technological Feedback" (in Milsum 1968, pp. 9–22).

The first extensive exposition of the idea of feedback loops can be found in a 1943 paper on "Behavior, Purpose, and Teleology," coauthored by Norbert Wiener, Julian Bigelow, and Arturo Rosenblueth.

10. For a full historical and theoretical perspective on this notion, see George P. Richardson, *Feedback Thought in Social Science and Systems Theory* (Philadelphia: University of Pennsylvania Press, 1992).

11. For readers unfamiliar with chaos and complexity theory, here I am including a brief history and description of some of its main concepts. For an extended, early account, see James Gleick, *Chaos: Making a New Science* (New York: Penguin Books, 1987).

Dissipative Structure. Ilya Prigogine, a Nobel laureate in physical chemistry, and his team at the University of Bruxelles have developed a nonlinear thermodynamics to describe spontaneous self-organization in open systems far from equilibrium. Whereas in classical thermodynamics dissipation of energy in heat transfer is always associated with waste or dissipation, in Prigogine's nonlinear version of thermodynamics dissipation becomes a source of order. He calls this phenomenon a stable "dissipative structure" that emerges in a state far from equilibrium, to distinguish it from the "equilibrium structure" (such as can be found in a crystal) of classical thermodynamics. Prigogine further demonstrates that dissipative structures can not only maintain themselves in a stable state far from equilibrium, but also evolve through fluctuations that are amplified by positive feedback loops. As the energy flowing through them increases, they may undergo new instabilities and turn into new structures of greater complexity. It is this insight of the constructive role of amplifying positive feedback loops that Prigogine contributes to cybernetics, which in its early mechanistic phase, or Cybernetics I, looked on such loops as destructive, rather than creative.

Autopoiesis. Humberto Maturana and Francisco Varela, among many other theorists of chaos and complexity, have adopted Prigogine's nonlinear description of self-organizing systems far from equilibrium as "order out of chaos." The two Chilean neuroscientists extend this systemic, nonlinear view to living organizations in their theory of "autopoiesis," a term meaning "self-making" and inspired by the concept of "poiesis" (which in ancient Greek means "making," or "creation") in Romantic literary theory. Maturana, the older of the two, began, at the University of Santiago de Chile, by studying the nervous system. He found it to be a closed network of mutually dependent interactions, in which any change in one part of a system affects the interrelations of all components in that system. In other words, Maturana discovered that the nervous system operates according to the principles of mutual causality. He advanced the hypothesis that the "circular organization" of the neurosystem corresponds to the basic organization of all living systems. Such systems are "organized in a closed causal circular process that allows for evolutionary change in the way the circularity is maintained, but not for the loss of circularity itself."

Like other systems theorists, Maturana points out that the activities of the nervous system cannot be separated from the environment in which it operates and therefore cannot be considered in terms of an external world entirely independent of the living organism. Cognition and perception do not "represent" or "reflect" an external reality, but select such a reality through the neural process of circular organization. In turn, this process is neither more nor less than the process of cognition itself. As Maturana puts it, "Living systems are cognitive systems, and living as a process is a process of cognition. This statement is valid for all organisms, with and without a nervous system" (Maturana and Varela 1980 [1970]).

Together with his former student Francisco Varela, Maturana proceeds to describe in more detail the notion of circular organization, based on mutual causality, for which they have invented a new term, "autopoiesis." Maturana and Varela define it as a system's network of processes of production, where each component participates in the generation and/or transformation of other components in that network. Thus, the network both produces and is, in turn, produced by its components. As Maturana and Varela have it, in a living system "the product of its operation is its own organization" (ibid., p. 82). The authors also make an important distinction between the structure and the organizational pattern of a living system. The structure is constituted by the set of actual interrelations of its components, i.e., the physical embodiment of its organization. The pattern of organization, on the other hand, refers to an abstract description of the system, independent of the features of its components.

Autopoiesis, then, is a pattern of organization common to all living systems, regardless of the nature and function of their components. Another significant characteristic of living systems is that through autopoiesis they also create a boundary that defines a particular system as a unit, circumscribing the field of operations of its network. So, living systems are both closed, in terms of their internal circuitry, and open, in terms of feedback loops with their environment. The openness and closedness of living systems are relative, however, ensuring a continuous dynamic of stability and change through negative and positive feedback loops.

Edge of Chaos. In turn, Stuart Kauffman (1991; 1993) has worked out the details of the nonlinear dynamics of stability and change in living systems, by applying the systems concept of nonlinear causality and Prigogine's notion of dissipative structures to evolutionary theory. Kauffman develops the concept of the "edge of chaos" to explain evolutionary changes at critical points of instability, far from equilibrium. Kauffman starts from the principle that in systems with perfect interior order, such as crystals, no change is possible. At the opposite end, in a chaotic system such as boiling water, there is hardly any order at all. So the system that will change the fastest must fall between the two extremes, on the "edge of chaos": it contains a certain amount of order, but with loosely interconnected components that can be altered with relative ease. As Kauffman points out, networks that are "on the boundary between order and chaos may have the flexibility to adapt rapidly and successfully through the accumulation of useful variations. In such poised systems, most mutations have small consequences because of the system's homeostatic nature. A few mutations, however, cause larger cascades of change. Poised systems will therefore typically adapt to a changing environment gradually, but if necessary, they can occasionally change rapidly" (Kauffman 1991).

Self-Organized Criticality and the Butterfly Effect. The theory of self-organized criticality assumes that small variations may, under certain conditions, trigger avalanches of change. This theory was first developed in mathematics, by René Thom, (1989 [1972]) who calls it "catastrophe theory," and then in physics by Per Bak (1996) who calls it "self-organized criticality." But its origins can equally be traced back to earlier general systems theory and cybernetics. Like the latter two, it employs nonlinear, mutual causality, assuming that very complicated, externally indecipherable patterns within a given system, although apparently "chaotic," can be determined by small, measurable changes inside the system. For example, Magoroh Maruyama (1968) writes that one of the advantages to be derived from scientific models based on mutual causality is that "simple rules may generate complex patterns, and that the patterns can greatly vary as the initial conditions vary" (in Milsum 1968, p. 100). Consequently, "the amount of 'information' required to describe the rules and the initial conditions is much smaller than the amount of 'information' needed to describe the whole range of variety of complex patterns" (ibid.).

Maruyama further notes that scientific models based on mutual causality require different research strategies. For example, in linear models, similar conditions are as a rule assumed to produce similar results, whereas dissimilar results are assumed to correspond to proportionally dissimilar conditions. So linear research has largely overlooked the possibility that mutual causal processes might amplify an insignificant, perhaps random, variation and consequently has often pursued a wrong track. According to Maruyama, even in a deterministic model, "the mutual causal process may disproportionately amplify an insignificant initial kick." In a probabilistic model, on the other hand, "a small deviation, which is within the range of high probability, may develop into a large deviation of very low probability (or more precisely, into a large deviation which is very improbable within the framework of probabilistic unidirectional causality). Therefore, in both a deterministic and a probabilistic universe, study of mutual causality circuitry may prove to be more fruitful than a study of initial conditions in many cases" (in Milsum 1968, p. 100). Here Maruyama describes what in chaos theory is known as the "butterfly effect" (Gleick 1987, pp. 20–23), according to which small changes in a system's initial state, such as a weather pattern, will lead over time to large-scale consequences (e.g., a butterfly's wings stirring the air in Paris today, can "cause" a storm in Shanghai a week later).

It is precisely this principle of nonlinear causality that lies at the foundation of Per Bak's notion of self-organized criticality (SOC). Bak (1996) describes what he calls a "canonical example of SOC," i.e., the sandpile behavior model. According to him, a pile of sand "exhibits punctuated equilibrium behavior, where periods of stasis are interrupted by intermittent sand slides. The sand slides, or avalanches, are caused by a domino effect, in which a single grain of sand pushes one or more other grains and causes them to topple. In turn, those grains of sand may interact with other grains in a chain reaction. Large avalanches, not gradual change, make the link between quantitative and qualitative behavior, and form the basis for emergent phenomena" (Bak 1996, p. 32). If one accepts the sandpile model as the way "nature works," then one must also accept "instability and catastrophe as inevitable in biology, history, and economics" (ibid.).

Furthermore, the same dynamics produces small ordinary events as it does large, catastrophic events. One must therefore abandon any illusion of long-term predictability or determinism in evolutionary biology, no less than in economics or history. Large events are contingent on a series of minor past events that are too complex to be condensed in terms of linear mathematical equations, the way mechanical events in physics might be. At most, one can explain why there is variability and change and what general patterns might emerge, that is, the systems' statistical features, but one can never predict what the particular outcome of a specific system or actual species will be. As Bak puts it, biological theory "will never predict elephants" (ibid., p. 10).

Punctuated Equilibrium. Bak offers the theory of punctuated equilibrium in evolution as a perfect example of self-organized criticality. This theory has been developed by Steven Jay Gould and Niles Eldredge, but its origins go at least as far back as Georges Cuvier's catastrophe theory. According to Gould and Eldredge (1977), evolution does not proceed in a steady, linear fashion, as Darwin believed, but is punctuated by sudden and unpredictable bursts of activity followed by long periods of stability. This view runs counter to habitual linear thinking, exhibited by evolutionary gradualists who look for specific external causes (such as the crash of a large meteorite on Earth leading to the extinction of dinosaurs) to explain cataclysmic occurrences. The evolutionary "bursts"—which can in any case be considered short only by geological standards, say, a million years of change, compared to six hundred million years of stasis—cannot be predicted any more than any other nonlinear phenomena, but they can be studied with the statistical methods of chaos and complexity theory as systems of self-organized criticality, far from equilibrium.

Strange Attractors and Fractals. Bak objects to the notion of "chaoplexity," however, preferring to keep the ideas of chaos and complexity theory separate. He argues that the simple chaotic systems, modeled on computers in experimental laboratories, "cannot produce a spatial fractal structure like the coast of Norway" (Bak 1996, p. 31). Although the topics of chaos and fractal geometry are mixed together again and again in popular literature, they "have little to do with each other." According to Bak, the confusion "arises from the fact that chaotic motion can be described in terms of mathematical objects known as strange attractors embedded in an abstract phase space. These strange attractors have fractal properties, but they do not represent geometrical fractals in real space like those we see in nature. In short, chaos theory cannot explain complexity" (ibid.).

Even though chaos theory cannot "explain" complexity, the link between the two concepts is nevertheless legitimate, precisely because of the way in which both use strange attractors. As Bak observes, phase space is different from "real space," in the sense that it is a mathematical abstraction, an artifice to measure the variables of a dynamic system, such as the swing of a pendulum. In this latter case, one can imagine a two-dimensional phase space, composed of two variables, velocity and angle, represented by a single point. As the imaginary pendulum swings back and forth, the point will move around a closed loop, so that the two variables will always be situated somewhere along this loop. Of course, in no sense can the loop be said to represent the actual trajectory of an actual pendulum ball. Rather, it is only a curve in an abstract mathematical space, composed of the system's two variables (velocity and angle). If one factors in friction, which will slow down and eventually stop the motion of the pendulum, one will represent this motion in phase space through a curve spiraling toward the center of the loop. Mathematicians call this spiraling trajectory an "attractor," because the fixed point at the center of the system can be said to "attract" the trajectory. They have also extended this metaphor to imaginary, frictionless pendulum trajectories.

The nonlinear mathematics of complexity distinguishes between three types of attractors: "periodic attractors" that correspond to periodic oscillations, like those of a frictionless pendulum swing; "point attractors" that correspond to systems reaching a stable equilibrium, such as a pendulum swing with friction factored in; and "strange attractors" that correspond to chaotic systems far from equilibrium. An example of the last type is the so-called "chaotic pendulum" of the Japanese mathematician Yoshisuke Ueda, where each swing is singular, so that each cycle covers a new area of phase space. In spite of this apparently irregular motion, the points in phase space form a complex, highly organized pattern that is now called the "Ueda strange attractor." In this respect, the term "chaos theory" can equally be seen as a misnomer, because nonlinear mathematicians distinguish between mere randomness or "noise" (i.e., what we laypersons commonly understand as chaos or disorder) and chaos as a deterministic, highly ordered pattern of strange attractors. As I pointed out in my discussion of the notion of chaos and disorder in chapter 1, nonlinear, physical–mathematical "chaos" and social or anthropological "chaos" are entirely different, nearly opposite concepts.

Yet, Benoît Mandelbrot's "fractal geometry" does provide an appropriate, nonlinear, mathematical language to describe the scale pattern of strange or "chaotic" attractors. In his influential book on *The Fractal Geometry of Nature* (French edition, 1975; English edition, 1983), the French mathematician proposed a nonlinear geometry to analyze the irregular shapes of natural phenomena, such as coastlines, clouds, mountains, lakes, rivers, lightning, etc. The distinguishing feature of these irregular, "fractal" shapes is that their patterns are "self-similar": they recur again and again at descending scales, so that their parts, at any scale, are similar in shape to the whole. Mandelbrot offers the obvious example of a cauliflower or a broccoli head, in which each smaller flower looks like the whole plant.

Originally, Mandelbrot was unaware of chaos theory and strange attractors, but he soon realized that such strange attractors were perfect examples of fractals. Nonlinear mathematicians now routinely describe strange attractors as trajectories in phase space that display fractal geometry. Finally, both fractal geometry and chaos theory have changed the mathematical idea of complexity. Linear mathematics as a rule describes simple structures by simple equations, and complicated structures by complicated equations. By contrast, in the nonlinear mathematics of complexity, simple equations may generate very complex strange attractors and simple iteration rules may produce extremely intricate patterns that can be modeled only on very large computers. Such computer-generated patterns have not only reinforced the idea of nature as symmetrical order in science, but have also been turned into a new art form.

Chapter 4

1. Capra himself has initiated this kind of intercultural comparative analysis in an earlier book, *The Tao of Physics: An Exploration of the Parallels between Modern Physics and Eastern Mysticism* (1975), which has over the years become a worldwide best seller.

2. See Joanna Macy, *Mutual Causality in Buddhism and General Systems Theory: The Dharma of Natural Systems* (Albany, N.Y.: State University of New York Press, 1991). In this groundbreaking book, Macy extends Capra's comparative analysis to systems theory, which she explores in relation to the early Buddhist view of reciprocal causality present in the Pali scriptures—the oldest documents to have recorded the teachings of Gautama Buddha and his closest disciples. Macy's study is a fine example of the kind of intercultural and cross-disciplinary research project that I have been pleading for in the present book.

3. In addition to Macy's study, David L. Hall and Roger T. Ames's *Thinking through Confucius* (1987) is another auspicious beginning for the kind of intercultural research that we need for developing global learning, conducive to global intelligence. This book is a fine, nonreductive and nonlinear, intercultural comparison of Western and Chinese ways of thinking and acting in the world, starting from the Confucian canon. For an intercultural comparative interpretation of Chinese Zen (Ch'an) Buddhism and Western postmodern thought, see Spariosu (1997, pp. 98–120).

4. As Hall and Ames point out, traditional Chinese polar thinking is different from Western dualistic or dichotomist thinking, precisely because the latter is linear, and the former is nonlinear or cyclical. A dualistic view of relationships such as can be found in various Western forms of dialectics leads to "an essentialistic interpretation in which the elements of the world are characterized by discreteness and independence." By comparison, a polar view, such as is to be found in early Taoism and Confucianism (as well as in early Buddhism),

involves "a contextualist interpretation of the world in which events are strictly interdependent" (Hall and Ames 1987, p. 19).

Hall and Ames further define polarity as a "relationship of two events each of which requires the other as a necessary condition for being what it is. Each existent is 'so of itself' and does not derive its meaning and order from any transcendent source. The notion of 'self' in the locution 'so of itself' has a polar relationship with 'other'. Each particular is a consequence of every other. And there is no contradiction in saying that each particular is both self-determinate and determined by every other particular, since each of the existing particulars is *constitutive* of every other as well. The principal distinguishing feature of polarity is that each pole can only be explained by reference to the other" (ibid., pp. 18–19; italics in the original). These descriptions are equally relevant to early Buddhist thinking. But, we shall see later on in this chapter that, properly speaking, Taoist thinking is not polar thinking any more than early Buddhism is, even though both may use this kind of thinking for their own strategic purposes.

5. For a good selection of Rumi's work, the reader may consult *Discourses of Rumi*, translated by A. J. Arberry (New York: Samuel Weiser, 1972) and Coleman Barks, *Essential Rumi* (San Francisco: Harper, 1997). For Rumi and Sufism, see W. C. Chittick, *The Sufi Path of Love: The Spiritual Teachings of Rumi* (Albany, N.Y.: State University of New York Press, 1983); and for Sufism in general, Chittick, *The Sufi Path of Knowledge* (Albany, N.Y.: State University of New York Press, 1989).

Chapter 5

1. See particularly the sections on "The Will to Power as Knowledge" (Nietzsche 1967, pp. 226–331) and "The Will to Power in Nature" (ibid., pp. 332–366).

2. The term "weighting principle" (Goodman 1978, p. 4) denotes the organizing or self-organizing principle(s) of any system, whether living or nonliving, according to which all components are ordered, interpreted, transformed, interlinked, and correlated to other living and nonliving systems. Examples of weighting principles are force, power, competition, domination, cooperation, peace, love, generosity, symbiosis, etc. Whereas most of the principles I have listed can be found in any living system, the weighting principle is what lends that system its specific structural and functional profile. For the purposes of the present book, "organizing" and "weighting" principles can be used interchangeably.

3. See Arne Naess's influential article, "The Shallow and the Deep, Long-Range Ecology Movement. A Summary," in *Inquiry*, vol. 16, 1973, pp. 95–100; and more recently, *Ecology, Community and Lifestyle*, trans. David Rothenberg (Cambridge: Cambridge University Press, 1989). An updated critical evaluation can be found in Gare (1993a, b; 1995).

4. For an incisive critique of the various aspects of contemporary ecological thinking, see Arran Gare (1993a), *Nihilism Incorporated*, especially chapter 2, "Responses to Environmental Problems." Unfortunately, Gare's post-Marxist solution to these problems does not go beyond "empowerment" and post-

Gramscian "hegemony." See my brief discussion of Gare's work, immediately below.

5. See *With a Fly's Eye, A Whale's Wit and a Woman's Heart*, edited by Theresa Corrigan and Stephanie Hoppe (San Francisco: Cleis Press, 1989).

6. For an extensive discussion of this topic, see Peter Singer, *Rethinking Life and Death: The Collapse of Our Traditional Ethics* (New York: St. Martin's Press, 1994). What Singer puts "in place of the old ethic" (pp. 187–222), however, is a version of enlightened utilitarianism that is hardly an "ethical revolution" (p. 187), unless we understand by revolution the return of the same, dressed in a new garb. Singer does not go beyond a mentality of violence and power, but only attempts to temper or "rationalize" it.

7. For a detailed discussion of the "ethopathology" of death in Western-style cultures, see Spariosu (1997, pp. 116–119).

Chapter 6

1. For a discussion of the republic of letters in the Renaissance and its use in subsequent efforts to reform the scholastic university, such as those of Vico in late seventeenth- and early eighteenth-century Naples, see Mazzotta (1999), *The New Map of the World*, especially chapter 2, "The Idea of a University," pp. 40–64.

2. For a comprehensive discussion of Newman's concept of the university, see Jaroslav Pelikan, *The Idea of the University: A Reexamination* (New Haven: Yale University Press, 1992).

3. Newman also anticipates and reduces *ad absurdum* such claims to intellectual unity as are based on an individual science or disciplinary field of knowledge. His remarks are worth quoting at length, being an early indictment of the misguided attempts at the "unity of knowledge" undertaken by reductionist scientists such as Wilson and Weinberg: "I say, let us imagine a project for organizing a system of scientific teaching [and, one can now add, research], in which the agency of man in the material world cannot allowably be recognized, and may allowably be denied. Physical and mechanical causes are exclusively treated of; volition is a forbidden subject. A prospectus is put out, with a list of sciences, we will say, Astronomy, Optics, Hydrostatics, Galvanism, Pneumatics, Statics, Dynamics, Pure Mathematics, Geology, Botany, Physiology, Anatomy, and so forth; but not a word about the mind and its powers, except what is said in explanation of the omission. That explanation is to the effect that the parties concerned in the undertaking have given long and anxious thought to the subject, and have been reluctantly driven to the conclusion that it is simply impracticable to include in the list of University lectures the Philosophy of Mind.... [Y]et, as time goes on, an omission which was originally but a matter of expedience, commends itself to the reason; and at length a professor is found, more hardy than his brethren, still however, as he himself maintains, with sincere respect for domestic feelings and good manners, who takes on him to deny psychology *in toto*, to pronounce the

influence of mind in the visible world a superstition, and to account for every effect which is found in the world by the operation of physical causes.... [O]ur Professor, I say, after speaking with the highest admiration of the human intellect, limits its independent action to the region of speculation, and denies that it can be a motive principle, or can exercise a special interference, in the material world.... At length he undertakes to show how the whole fabric of material civilization has arisen from the constructive powers of physical elements and physical laws" (Newman 1947, pp. 49–51). Here Newman uncannily anticipated, through what to him seemed an unlikely, if not absurd, thought experiment, the whole subsequent history of twentieth-century scientific reductionism.

4. For brilliant analyses of the current woes of the American universities and their bankrupt principle of "academic excellence," see, among others, Bruce Wilshire, *The Moral Collapse of the University: Professionalism, Purity and Alienation* (Albany, N.Y., 1990) and William Readings, *The University in Ruins* (Cambridge, Mass., 1996). An earlier, excellent study is Jacques Barzun, *The American University: How It Runs, Where It Is Going* (London, 1969). For the university and the humanities, see Spellmeyer, *Arts of Living* (2003).

5. For a full discussion of the canon wars in contemporary North American academia see my *Remapping Knowledge*, chapter 2.

6. Cited in Edward Danforth Eddy, Jr., *Colleges for Our Land and Time* (New York: Harper and Brothers, 1957) p. 33.

Chapter 7

1. I am grateful to Mikhail Epstein of Emory University for this proposal, which he originally presented at the Concord Retreat on "The Future of the Humanities" (January 2000) and which I have adapted and substantially modified here. For a fully developed curriculum in intercultural studies, based on the principles that I have outlined in this book, see the last chapter of my *Remapping Knowledge.*

References

Abedi, Mehdi, and Michael Fischer. 1990. *Debating Muslims: Cultural Dialogues in Postmodernity and Tradition*. Madison: University of Wisconsin Press.

Abdullah, Sharif. 1999. *Creating a World That Works for All*. San Francisco: Berrett-Koehler.

Abu-Lughod, Janet L. 1989. *Before European Hegemony: The World System A.D. 1250–1350*. New York: Oxford University Press.

Aczel, Richard, and Peter Nemes, editors. 2003. *The Finer Grain*. Bloomington, Ind.: Indiana University Press.

American Council on Education (ACE). 2002. "Beyond September 11: A Comprehensive National Policy on International Education." Available at ⟨http://www.acenet.edu/bookstore⟩.

Anderson, Benedict. 1991. *Imagined Communities: Reflections on the Origin and Spread of Nationalism*. Revised edition. London and New York: Verso.

Aṅguttara Nikāya. 1999. Translated as *Numerical Discourses of the Buddha: An Anthology of Suttas from the Aṅguttara Nikāya*. Edited and translated by Nyanaponika Thera and Bhikkhu Bodhi. Walnut Creek, Calif.: AltaMira Press.

Appadurai, Arjun. 1996. *Modernity at Large: Cultural Dimensions of Globalization*. Chicago: Chicago University Press.

Ashby, Eric. 1966. *Universities: British, Indian, African: A Study in the Ecology of Higher Education*. London: Weidenfeld and Nicolson.

Bak, Per. 1996. *How Nature Works*. New York: Springer-Verlag New York.

Bakhtin, Mikhail. 1981. *The Dialogic Imagination: Four Essays*. Edited by Michael Holquist and translated by Caryl Emerson and Michael Holquist. Austin: University of Texas Press.

Barber, Benjamin R. 1995. *Jihad vs. McWorld: How Globalism and Tribalism Are Reshaping the World*. New York: Ballentine Books.

Barnet, Richard J., and John Cavanaugh. 1994. *Global Dreams: Imperial Corporations and the New World Order*. New York: Simon and Schuster.

Barzun, Jacques. 1969. *The American University: How It Runs, Where It Is Going*. London: Oxford University Press.

Bauman, Zygmunt. 2000. *Globalization: The Human Consequences*. New York: Columbia University Press.

Bateson, Gregory. 2000 [1972]. *Steps to an Ecology of Mind*. Chicago: University of Chicago Press.

Benyus, J. M. 1997. *Biomimicry: Innovations Inspired by Nature*. New York: William Morrow.

Berger, Peter, and Samuel P. Huntington, editors. 2002. *Many Globalizations: Cultural Diversity in the Contemporary World*. New York: Oxford University Press.

Berry, Wendell. 2001. *Life Is a Miracle: An Essay against Modern Superstition*. New York: Counterpoint Press.

Boulding, Elise. 1988. *Toward a Global Civic Culture: Education for an Interdependent World*. New York: Teachers College Press, Columbia University.

Boulding, Kenneth E. 1978. *Ecodynamics: A New Theory of Societal Evolution*. London: Sage.

———, and Norman Meyers. 1990. *The Gaia Atlas of Future Worlds: Challenge and Opportunity in an Age of Change*. London: Gaia Books.

Bourdieu, Pierre. 1998. *Contre-feux: Propos pour servir à la résistance contre l'invasion néo-libérale*. Paris: Éditions Liber.

Bruyn, Severyn. 2000. *A Civil Economy: Transforming the Marketplace in the 21st Century*. Ann Arbor: University of Michigan Press.

Capra, Fritjof. 1997. *The Web of Life: A New Synthesis of Mind and Matter*. London: Harper Collins.

———. 1975. *The Tao of Physics: An Exploration of the Parallels between Modern Physics and Eastern Mysticism*. London: Harper Collins.

Castells, Manuel. 1996. *The Rise of the Network Society*. Oxford: Blackwell.

———. 1997. *The Power of Identity*. Oxford: Blackwell.

———. 1998. *End of Millennium*. Oxford: Blackwell.

Chittick, W. C. 1989. *The Sufi Path of Knowledge*. Albany, N.Y.: State University of New York Press.

———. 1983. *The Sufi Path of Love: The Spiritual Teachings of Rumi*. Albany, N.Y.: State University of New York Press.

Chuang Tzu. 1997. *The Inner Chapters*. Translated by David Hinton. Washington D.C.: Counterpoint.

Confucius. 1979. *The Analects*. Translated by D. C. Lau. London: Penguin Books.

Corrigan, Theresa, and Stephanie Hoppe. 1989. *With a Fly's Eye, a Whale's Wit and a Woman's Heart*. San Francisco: Cleis Press.

Costa, Ruy O. 2003. "Globalization and Culture." Report to the Advisory Committee on Social Witness Policy, U.S. Presbyterian Church. Louisville: Presbyterian Distribution Center.

Currie, Janice K., and Janice Newson, editors. 1998. *Universities and Globalization: Critical Perspectives*. London: Sage.

Cvetkovich, Ann, and Douglas Kellner, editors. 1997. *Articulating the Global and the Local*. Boulder, Colo.: Westview Press.

Daly, Herman. 1996. *Beyond Growth: The Economics of Sustainable Development*. Boston: Beacon Press.

Dimock, Wai Chee. 1997. "A Theory of Resonance." *PMLA* 112, 5 (October): 1061–1071.

Dyson, George B. 1997. *Darwin among the Machines: The Evolution of Global Intelligence*. New York: Addison–Wesley.

Eagleton, Terry. 2000. *The Idea of Culture*. Oxford: Blackwell.

Eddy, Jr., Danforth Edward. 1957. *Colleges for Our Land and Time*. New York: Harper and Brothers.

Eisler, Riane. 1988. *The Chalice and the Blade: Our History, Our Future*. New York: Harper Collins.

Etzioni, Amitai. 1988. *The Moral Dimension: Toward a New Economics*. New York and London: The Free Press.

Even-Zohar, Itamar. 1990. *Polysystem Studies*. Special issue of *Poetics Today* 11: 1–2.

Featherstone, Mike. 1995. *Undoing Culture: Globalization, Postmodernism, and Identity*. London: Sage.

Foucault, Michel. 1975. *Surveiller et punir*. Paris: Gallimard.

Frager, Robert, and James Fadiman, editors. 1997. *Essential Sufism*. Edison, N.J.: Castle Books.

Friedman, Thomas. 2000. *The Lexus and the Olive Tree: Understanding Globalization*. New York: Farrar, Straus and Giroux.

Fukuyama, Francis. 1992. *The End of History and the Last Man*. London: Penguin Books.

Gare, Arran E. 1993a. *Nihilism Incorporated: European Civilization and Environmental Destruction*. Bungadore, Australia: Eco-Logical Press.

———. 1993b. *Beyond European Civilization: Marxism, Process Philosophy, and the Environment*. Bungadore, Australia: Eco-Logical Press.

———. 1995. *Postmodernism and Environmental Crisis*. London and New York: Routledge.

Giddens, Anthony. 1999. *Runaway World: How Globalization Is Reshaping Our Lives*. London: Profile Books.

Girard, René. 1977. *Violence and the Sacred*. Baltimore, Md.: Johns Hopkins University Press.

———. 1986. *The Scapegoat*. Baltimore, Md.: Johns Hopkins University Press.

———. 1987. *Things Hidden since the Foundation of the World*. Stanford, Calif.: Stanford University Press.

Gleick, James. 1987. *Chaos: Making a New Science*. New York: Penguin Books.

Goodman, Nelson. 1978. *Ways of Worldmaking*. Indianapolis, Ind.: Hackett.

Gorbachev, Mikhail S. 1991. *A Road to the Future*. Santa Fe, N.M.: Ocean Tree Books.

———, and Zdenek Mlynar. 2003. *Conversations with Gorbachev*. New York: Columbia University Press.

Gould, Stephen Jay, and Niles Eldredge. 1977. "Punctuated Equilibrium: The Tempo and Mode of Evolution Reconsidered." *Paleobiology* 3: 115–151.

Gray, John. 1999. *False Dawn: The Delusions of Global Capitalism*. London: Granta Books.

Hall, David L., and Roger T. Ames. 1987. *Thinking through Confucius*. Albany, N.Y.: State University of New York Press.

Hardt, Michael, and Antonio Negri. 2000. *Empire*. Cambridge, Mass.: Harvard University Press.

Harrison, Lawrence E., and Samuel P. Huntington, editors. 2000. *Culture Matters: How Values Shape Human Progress*. New York: Basic Books.

Hawken, Paul. 1993. *The Ecology of Commerce: A Declaration of Sustainability*. New York: Harper Collins.

Held, David. 1995. *Democracy and the Global Order: From the Modern State to Cosmopolitan Governance*. Cambridge: Polity Press.

Henderson, Hazel. 1978. *Creating Alternative Futures: The End of Economics*. West Hartford, Conn.: Kumarian Press.

———. 1999. *Beyond Globalization: Shaping a Sustainable Global Economy*. West Hartford, Conn.: Kumarian Press.

Holling, C. S. 1995. "What Barriers? What Bridges?" In Lance H. Gunderson, C. S. Holling, and Stephen S. Light, eds., *Barriers and Bridges to the Renewal of Ecosystems and Institutions*. New York: Columbia University Press.

Huntington, Samuel P. 1996. *The Clash of Civilizations and the Remaking of World Order*. New York: Simon and Schuster.

Jameson, Fredric, and Masao Miyoshi, editors. 1998. *The Cultures of Globalization*. Durham, N.C.: Duke University Press.

Jammer, Max. 1957. *Concepts of Force: A Study in the Foundations of Dynamics*. Cambridge, Mass.: Harvard University Press.

Jowitt, Kenneth. 1993. *The New World Disorder: The Leninist Extinction*. Berkeley: University of California Press.

Kahn, Herman. 1982. *The Coming Boom: Economic, Political and Social*. New York: Simon and Schuster.

———, William Brown, and Leon Martel. 1977. *The Next 200 Years: A Scenario for America and the World*. London: Associated Business Programs.

Kahn, Herman, and Julian L. Simon, editors. 1984. *The Resourceful Earth: A Response to Global 2000*. Oxford: Blackwell.

Kauffman, Stuart. 1991. "Antichaos and Adaptation." *Scientific American* (August): 78–84.

———. 1993. *The Origins of Order: Self-Organization and Selection in Evolution*. New York: Oxford University Press.

Kavanaugh, John F. 1991a. *Following Christ in a Consumer Society: The Spirituality of Cultural Resistance*. Maryknoll, N.Y.: Orbis Books.

Kavanaugh, John F., and Mev Puleo. 1991b. *Faces of Poverty, Faces of Christ*. Maryknoll, N.Y.: Orbis Books.

King, Anthony D., editor. 1997. *Culture, Globalization, and the World System: Contemporary Conditions for the Representation of Identity*. Minneapolis: University of Minnesota Press.

Kuhn, Thomas. 1970 [1962]. *The Structure of Scientific Revolutions*. Second revised edition. Chicago: Chicago University Press.

Lao Tzu. 1963. *Tao Te Ching*. Translated by D. C. Lau. London: Penguin Books.

Lash, Scott, and John Urry, editors. 1988. *The End of Organized Capitalism*. Boulder, Colo.: Lightning Source.

Lash, Scott, Mike Featherstone, and Roland Robertson, editors. 1995. *Global Modernities: From Modernism to Hypermodernism and Beyond*. London: Sage.

Laszlo, Ervin. 1969. *System, Structure, Experience*. New York: Gordon and Breach.

———. 1973. *Introduction to Systems Philosophy*. New York: Harper.

———. 1974. *A Strategy for the Future: The Systems Approach to World Order*. New York: George Braziller.

———. 2003. *You Can Change the World: The Global Citizen's Handbook for Living on Planet Earth: A Report of the Club of Budapest*. Kansas City: Midpoint Trade Books.

Loewe, Michael. 1982. *Chinese Ideas of Life and Death*. London: Allen and Unwin.

Lovelock, James. 1988. *The Ages of Gaia: A Biography of Our Living Earth*. New York: W.W. Norton.

Lovins, Amory, L. Hunter Lovins, and Paul Hawken. 1999. *Natural Capitalism: Creating the Next Industrial Revolution*. Boston: Little, Brown.

Loye, David. 2002. *Darwin's Lost Theory*. New York: Seven Stories Press.

Macy, Joanna. 1991. *Mutual Causality in Buddhism and General Systems Theory: The Dharma of Natural Systems*. Albany, N.Y.: State University of New York Press.

Mandelbrot, Benoît. 1983 [1975]. *The Fractal Geometry of Nature*. New York: Freeman.

Marcus, Steven, and Michael Fischer. 2000. *Anthropology As Cultural Critique: An Experimental Moment in the Human Sciences*. Second edition. Chicago: University of Chicago Press.

Margulis, Lynn. 1989. "Gaia: The Living Earth." Dialogue with Fritjof Capra. *The Elmwood Newsletter*, Berkeley, Calif., vol. 5, no. 2.

———. 1992. *Diversity of Life*. Berkeley Heights, N.J.: Enslow.

Margulis, Lynn, and Dorion Sagan. 1986. *Microcosm*. New York: Summit.

———. 1995. *What Is Life?* New York: Simon and Schuster.

Maruyama, Magoroh. 1974. "Paradigmatology and Its Application to Cross-disciplinary, Cross-professional and Cross-cultural Communication." *Cybernetica* 17: 135–156, 237–286.

Mathiesen, Thomas. 2001. *Prison on Trial: A Critical Assessment*. Winchester: Waterside Press.

Maturana, Humberto, and Francisco Varela. 1980. *Autopoiesis and Cognition*. Dordecht: Reidel.

Mazzotta, Giuseppe. 1999. *The New Map of the World: The Poetic Philosophy of Giambattista Vico*. Princeton, N.J.: Princeton University Press.

McLuhan, Marshall, and Quentin Fiore. 1968. *War and Peace in the Global Village*. New York: Bantam.

Meadows, Donella H., Dennis Meadows, and Jørgen Randers. 1972. *The Limits to Growth*. New York: Universe Books.

———. 1993. *Beyond the Limits: Confronting Global Collapse, Envisioning a Sustainable Future*. Post Mills, Vt.: Chelsea Green Publishing.

Melko, Matthew. 1969. *The Nature of Civilizations*. Boston: Porter Sargent.

Merezhkovsky, Konstantin S. 1909. *Theory of Symbiogenesis* (in Russian). Kazan.

Milsum, John H., editor. 1968. *Positive Feedback: A General Systems Approach to Positive/Negative Feedback and Mutual Causality*. London: Pergamon Press.

Myrdal, Gunnar. 1957. *Economic Theory and Underdeveloped Regions*. London: Duckworth.

Naess, Arne. 1973. "The Shallow and the Deep, Long-Range Ecology Movement. A Summary." *Inquiry* 16: 95–100.

———. 1989. *Ecology, Community, and Lifestyle*. Translator David Rothenberg. Cambridge: Cambridge University Press.

Needham, Joseph. 1978. *The Shorter Science and Civilization in China: An Abridgement of Joseph Needham's Original Text*. Volume 1. Prepared by Colin A. Ronan. New York: Cambridge University Press.

Newman, John Henry Cardinal. 1947 [1873]. *The Idea of a University*. Edited with a Preface and Introduction by Charles Frederick Harrold. New York: Longmans, Green.

Nietzsche, Friedrich. 1956 [1887]. *The Genealogy of Morals: An Attack*. Translated by Francis Golffing. New York: Vintage.

———. 1967 [1901]. *The Will to Power*. Edited by Walter Kaufmann. Translated by Walter Kaufmann and R. J. Hollingdale. New York: Vintage.

Nye, Joseph. 1990. *Bound to Lead: The Changing Nature of American Power*. New York: Basic Books.

———. 2004. *Soft Power: The Means to Success in World Politics*. Boulder, Colo.: Public Affairs.

Orr, David W. 1993. *Ecological Literacy: Education and the Transition to a Postmodern World*. Albany, N.Y.: State University of New York Press.

Orwell, George. 1949. *Nineteen Eighty-Four*. London: Secker and Warburg.

Payne, James. 1989. *Why Nations Arm*. Oxford: Blackwell.

Pelikan, Jaroslav. 1992. *The Idea of the University: A Reexamination*. New Haven: Yale University Press.

Pratt, Mary Louise. 1992. *Imperial Eyes: Travel Writing and Transculturation*. New York: Routledge.

Prigogine, Ilya, and Isabelle Stengers. 1984. *Order Out of Chaos*. New York: Bantam.

Readings, William. 1996. *The University in Ruins*. Cambridge, Mass.: Harvard University Press.

Reich, Robert. 1991. *The Work of Nations: Preparing Ourselves for the 21st Century*. New York: Knopf.

———. 2001. *The Future of Success*. New York: Heinemann.

Richardson, George P. 1992. *Feedback Thought in Social Science and Systems Theory*. Philadelphia: University of Pennsylvania Press.

Robertson, Roland. 1992. *Globalization: Social Theory and Global Culture*. London: Sage.

Rosenblueth, Arturo, Norbert Wiener, and Julian Bigelow. 1943. "Behavior, Purpose, and Teleology." *Philosophy of Science* 10: 18–24.

Roszak, Theodore. 1975. *Unfinished Animal: The Aquarian Frontier and the Evolution of Consciousness*. London: Faber and Faber.

Rumi, Jalal al-Din. 1961. *Discourses of Rumi*. Translated by A. J. Arberry. Reprint, New York: Samuel Weiser, 1972.

———. 1997. *Essential Rumi*. Translated by Coleman Barks. San Francisco: Harper.

Saṃyutta Nikāya: The Connected Discourses of the Buddha. 2002. Translated by Bhikkhu Bodhi. Boston: Wisdom Publications.

Schumacher, E. F. 1975. *Small Is Beautiful: Economics As If People Mattered*. New York: Harper and Row.

Seed, John, Pat Fleming, Joanna Macy, and Arne Naess. 1988. *Thinking Like a Mountain: Toward a Council of All Beings*. Philadelphia: New Society Publishers.

Sen, Amartya. 1987. *On Ethics and Economics*. Oxford: Blackwell.

———. 2000. *Development as Freedom*. New York: Anchor Books.

Sidgwick, Henry. 1874. *The Methods of Ethics*. London: Macmillan.

Singer, Peter. 1985. *In Defense of Animals*. New York: Basil Blackwell.

———. 1994. *Rethinking Life and Death: The Collapse of Our Traditional Ethics*. New York: St. Martin's Press.

———. 1995. *How Are We To Live? Ethics in an Age of Self-Interest*. Amherst, N.Y.: Prometheus Books.

Sloterdijk, Peter. 1998. *Sphaeren I. Blasen*. Frankfurt am Main: Suhrkampf Verlag.

Smith, John Maynard. 1975. *The Theory of Evolution*, third edition. New York: Cambridge University Press.

Soros, George. 1998. *The Crisis of Global Capitalism*. New York: Public Affairs.

Spariosu, Mihai I. 1989. *Dionysus Reborn: Play and the Aesthetic Dimension in Modern Philosophical and Scientific Discourse*. Ithaca, N.Y.: Cornell University Press.

———. 1991. *God of Many Names: Play, Poetry, and Power in Hellenic Thought from Homer to Aristotle*. Durham, N.C.: Duke University Press.

———. 1997. *The Wreath of Wild Olive: Play, Liminality, and the Study of Literature*. Albany, N.Y.: State University of New York Press.

———. 2003. "Global Intelligence and Intercultural Studies." *Turnrow* 3, 1 (fall): 102–126.

———. 2005. *Remapping Knowledge: Intercultural Studies for a Global Age*. New York and Oxford: Berghahn.

Spengler, Oswald. 1926. *The Decline of the West*. New York: Oxford University Press.

Stock, Gregory. 1993. *Metaman*. London: Bantam Press.

Stone, Lawrence, editor. 1974. *The University in Society*. Volume 1. Princeton, N.J.: Princeton University Press.

———. 1975. *The University in Society*. Volume 2. Princeton, N.J.: Princeton University Press.

Streng, Frederick. 1975. "Reflections on the Attention Given to Mental Constructions in the Indian Buddhist Analysis of Causality." *Philosophy East and West* 25 (January), 1: 71–80.

Symonides, Janusz, and Kishore Singh et al. 1996. *From a Culture of Violence to a Culture of Peace*. Paris: UNESCO.

Thom, René. 1989 [1972]. *Structural Stability and Morphogenesis: An Outline of a General Theory of Models*. New York: Perseus Publishing.

Tomlinson, John. 1999. *Globalization and Culture*. Chicago: University of Chicago Press.

Toynbee, Arnold. 1934–1961. *A Study of History*. 12 vols. New York: Oxford University Press.

Trungpa, Chögyam. 1987. Commentary to the *Tibetan Book of the Dead: The Great Liberation through Hearing in the Bardo*. Boston: Shambala.

Udāna. 1935. Translated by F. L. Woodward. Sacred Books of the Buddhists, vol. VIII. London: The Pali Text Society.

Vaihinger, Hans. 1924. *The Philosophy of "As If": A System of the Theoretical, Practical, and Religious Fictions of Mankind*. Translated by C. K. Ogden. London: Macmillan.

Vico, Giambattista. 1968 [1744]. *The New Science of Giambattista Vico*. Translated by Max H. Fisch and Thomas G. Bergin. Ithaca, N.Y.: Cornell University Press.

Von Bertalanffy, Ludwig. 1968. *General Systems Theory*. New York: George Braziller.

Wallerstein, Immanuel. 1991. *Geopolitics and Geoculture: Essays on the Changing World-system*. Cambridge: Cambridge University Press.

———. 1999. *The End of the World As We Know It: Social Science for the Twenty-First Century*. Minneapolis: University of Minnesota Press.

Weinberg, Steven. 1992. *Dreams of a Final Theory: The Scientist's Search for the Ultimate Laws of Nature*. New York: Vintage Books.

Wells, H. G. 1994 [1938]. *World Brain: H. G. Wells on the Future of World Education*. London: Adamantine Press.

Whewell, William. 1995 [1840]. *The Philosophy of the Inductive Sciences*. In *Collected Works of William Whewell*. Edited by Richard Yeo. Chicago: Chicago University Press.

Wiener, Norbert. 1965 [1948]. *Cybernetics: The Science of Control and Communication in the Animal and the Machine*. Second edition. Cambridge, Mass.: The MIT Press.

Wilde, Oscar. 1954. *The Works of Oscar Wilde*. Edited by G. F. Maine. New York: E.F. Dutton.

Wilshire, Bruce. 1990. *The Moral Collapse of the University: Professionalism, Purity, and Alienation*. Albany, N.Y.: State University of New York Press.

Wilson, Edward O. 1998. *Consilience: The Unity of Knowledge*. London: Little, Brown.

———. 2003. *The Future of Life*. New York: Knopf.

Index